Pollution Evaluation

THE QUANTITATIVE ASPECTS

ENVIRONMENTAL SCIENCE AND TECHNOLOGY SERIES

Series Editor

J. Carrell Morris

Gordan McKay Professor of Sanitary Engineering
Harvard University
Cambridge, Massachusetts

Additional Volumes in Preparation

Pollution Evaluation

THE QUANTITATIVE ASPECTS

William F. Pickering

Professor of Chemistry
The University of Newcastle
New South Wales, Australia

Marcel Dekker, Inc.   New York and Basel

Library of Congress Cataloging in Publication Data

Pickering, William F
   Pollution evaluation.

   (Environmental science and technology series ; v. 2)
   Includes index.
   1. Pollution--Measurement. 2. Environmental
chemistry. 3. Chemistry, Analytic. I. Title.
II. Series.
TD177.P5   628.5   77-12910
ISBN 0-8247-6621-0

MARCEL DEKKER, INC.
270 Madison Avenue, New York, New York 10016

Current printing (last digit):
10 9 8 7 6 5 4 3 2 1

PRINTED IN THE UNITED STATES OF AMERICA

PREFACE

This book has been prepared in response to several seemingly unrelated modern developments.

In an era when quantitative measurements are being quoted by a wider cross section of the community, there is a need for more people to evaluate critically the significance of the numbers released in technical or semitechnical publications. The ideal basis for such critical assessment is a sound training in the principles of physical and chemical analysis.

Unfortunately, people who are interested in the social science aspects of pollution and environmental matters rarely have the opportunity, or time, to undertake a rigorous course. At the same time, general chemistry students complain about the lack of relevance of much of the material presented, and courses in quantitative analysis are often just further hurdles in the path to the qualification sought at the end of the study period.

Trace analysis (and thus pollution evaluation) can be an excellent medium for inculcating the philosophy of the scientific method, for developing the power of critical evaluation, and for building a bridge between analytic principles and socially relevant problems.

This book has therefore been developed on a dual theme. The odd chapters discuss, in fairly general terms, modes of evaluating some typical forms of environmental pollution, with comments on some of the natural factors requiring consideration. Methods of evaluation are improving continuously, hence the procedures quoted are illustrative, rather than representing the approach of any particular environmental control agency.

The even chapters take up the fundamental principles of some of the techniques quoted in the discussion of pollution evaluation. For the sake of brevity, these chapters necessarily omit much of the rigor considered desirable by the dedicated teacher of chemical analysis. This problem is not insurmountable, for if the relevance of the applications provides sufficient interest and motivation, one can augment this introductory approach with material from any of the high-quality analytic texts now available.

The objectives of this author will be met if the material contained within this publication causes prospective environmental experts to make critical evaluations of the significance of the numbers produced in laboratories, and simultaneously causes routine analysts to consider factors arising outside the walls of the testing laboratory. As a generation, we are concerned with pollution; as scientists we should be concerned with the validity of the mass of data being generated; as people we should maintain a balanced approach and not be misled by personal bias, unfounded emotionalism, or distorted perspectives.

Pollution evaluation is a complex problem involving many arbitrary decisions; quantitative chemical analysis is an exact science. The bridges between have to be built with care.

The idea or seed for this publication was planted during the stimulating period the author was attached (as a Senior Killam Fellow) to the Trace Analysis Center, Dalhousie University, Halifax. It germinated during visits to various research centers in different parts of the western world, and has been nurtured by comments and discussions with many colleagues.

Permission to reproduce diagrams was provided willingly by Elsevier and Marcel Dekker, and the Department of Mechanical Engineering, University of Newcastle, kindly made available local air pollution data.

My scribbled notes were transcribed efficiently into a quality presentation by Mrs. E. K. Swift and Miss K. Manthey, and my wife shared in the chore of proofreading.

To all who have directly, or unwittingly, aided, abetted, or encouraged, I offer a sincere thank you.

WILLIAM F. PICKERING

CONTENTS

INTRODUCTION TO POLLUTION EVALUATION

I. LEGAL ASPECTS OF CHEMICAL SCREENING PROCESSES

In recent years, increased public awareness of the toxic effects of many chemical compounds has created strong social pressures with respect to the protection of the environment and human health. One side effect of this has been a proliferation of governmental control measures, many of which quote permitted concentration limits with concomitant penalties for breaches of regulations. In the same era, the cost of routine control (i.e., screening) of manufactured products has been minimized by the introduction of automated analytical instruments operated by technicians, and this trend has been avidly adopted by many agencies responsible for monitoring the environment and the clinical status of the population. However, since process control is predominantly an internal quality-matching procedure, some consistent errors can be tolerated. On the other hand, as soon as legal limits are prescribed, one has to ensure that the data produced in laboratories is of sufficient quality to withstand courtroom challenges.

The accuracy of chemical determinations depends on both the skill of the analyst and the procedure or technique adopted, and in some regulations it is considered desirable to specify the methods of analysis to be used. Unfortunately, some of the chemical tests listed are outmoded, and only rarely is mention made of approved sampling procedures or limitations, such as ultimate sensitivity, range of applicability, potential interferants, degree of accuracy achievable, etc. In most regulations the term "chemist" does not appear, nor is there reference to certification of the chemical analysis. Thus, theoretically, unskilled labor could be used to produce the numbers placed on various reports.

The latter aspect is bound to be amended as increasing numbers of industries appeal against pollution fines, or if citizens seek to claim damages for faulty medical diagnoses based on erroneous clinical screening information.

The actual task of selecting appropriate measuring procedures, and tabulating justifiable legal limits, is a massive one and should involve large numbers of experienced chemists working in conjunction with other professions.

The legal minded are concerned mainly with defining or determining what evidence under the bill, act, or regulations is acceptable to the courts, and the status of the person producing the data is of secondary consideration unless the regulatory document specifically defines a role for professionals.

A defendant, on the other hand, may choose to create doubt about the validity of the quantitative evidence being used against him. In such cases, the questions to be answered include:

Was the sample truly representative?

Were duplicate samples forwarded to a second, independent laboratory for quantitative study?

What are the limits of detection, accuracy, degree of reliability, etc., of the analytic procedure used?

Who undertook the analyses, a semiskilled operator, a qualified technician, or a professional chemist?

What precautions were taken to avoid contamination of the sample during handling?

Does the mode of reporting (e.g., % of element) reflect the real toxic situation? In other words, does the health hazard depend on the amount of a particular element present or on the existence of a particular chemical form?

As the complexity of chemical analysis increases (particularly with respect to evaluation of trace amounts of molecular species), the evidence of expert witnesses will become a greater feature of litigation.

Table 1.1 summarizes a few areas where legislative controls could result in enhanced status for chemists. Each area will provide many new scientific challenges in the decades to come, and this should ensure an increased demand for highly qualified analytical chemists. At the same time, there is an onus on the nonspecialist to become familiar with the difficulties associated with producing valid chemical analyses. No decision or judgment can be better than the data on which it is based.

Table 1.1

Areas of Scientific Challenge in Which Legislative Controls
may Enhance the Status of Chemists

Involved
Groups:

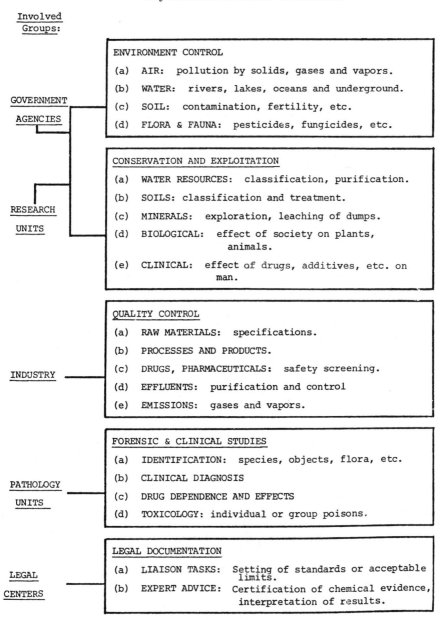

GOVERNMENT

AGENCIES

ENVIRONMENT CONTROL

(a) AIR: pollution by solids, gases and vapors.

(b) WATER: rivers, lakes, oceans and underground.

(c) SOIL: contamination, fertility, etc.

(d) FLORA & FAUNA: pesticides, fungicides, etc.

RESEARCH

UNITS

CONSERVATION AND EXPLOITATION

(a) WATER RESOURCES: classification, purification.

(b) SOILS: classification and treatment.

(c) MINERALS: exploration, leaching of dumps.

(d) BIOLOGICAL: effect of society on plants,
                 animals.

(e) CLINICAL: effect of drugs, additives, etc. on
                man.

INDUSTRY

QUALITY CONTROL

(a) RAW MATERIALS: specifications.

(b) PROCESSES AND PRODUCTS.

(c) DRUGS, PHARMACEUTICALS: safety screening.

(d) EFFLUENTS: purification and control

(e) EMISSIONS: gases and vapors.

PATHOLOGY

UNITS

FORENSIC & CLINICAL STUDIES

(a) IDENTIFICATION: species, objects, flora, etc.

(b) CLINICAL DIAGNOSIS

(c) DRUG DEPENDENCE AND EFFECTS

(d) TOXICOLOGY: individual or group poisons.

LEGAL

CENTERS

LEGAL DOCUMENTATION

(a) LIAISON TASKS: Setting of standards or acceptable
                     limits.

(b) EXPERT ADVICE: Certification of chemical evidence,
                     interpretation of results.

Even if legal considerations can be ignored, there is a moral obligation to understand the significance of any numbers produced by sophisticated equipment or reported by remote testing authorities.

It is hoped that the process of interleaving a general discussion of pollution evaluation, with brief comments on fundamental principles of chemical analysis, will produce a mental outlook appropriate to wise decision making

II. VALIDITY OF DATA

A. Analytical Problems

Despite the intense efforts made in recent years, our knowledge of the modes of distribution of trace elements within the biosphere and lithosphere is still in a comparatively primitive state. There is a paucity of information on the nature of the most significant chemical forms, and it is frequently difficult to accept the validity of published numbers because of poor control of some, most, or all aspects of the analytical process. One cynic has remarked that often it would be equally appropriate to "pick a number, any number."

The challenges associated with developing procedures for valid monitoring of polluted systems can be subdivided into four major categories:

1. Development of techniques that are sufficiently sensitive, i.e., allow accurate evaluation of components present at part per billion (i.e., μg/liter) levels

2. Evaluation of errors associated with sample collection, storage, and preparation for testing

3. Definition of type of chemical analysis required

4. Designation of the role of competing equilibria

There have been major advances in the area of new techniques, but the significance of the other facets is only currently being appreciated.

Some aspects of the sampling problem are considered in the next chapter, and several of the analytic techniques that have proved sufficiently sensitive for pollution studies are discussed in the analytical techniques segments of Chapters 4, 6, and 8. Accordingly, at this stage, discussion will be directed primarily to the two remaining facets, the selection of appropriate marker systems and the role of competing equilibria.

B. Selection of Marker Species

One difficulty facing all investigators is clear definition of the type of data required. Is the determination of the elemental composition sufficient, or should one be ascertaining the amount of some particular molecular species or functional group?

Analysis for the total amount of some given element (e.g., Hg, Pb, P) can yield an incomplete or erroneous measure of potential health hazards. Consider the relationship between the total metal content of soils and uptake by plants.

Only a fraction of the total soil metal content is available to plants, and the investigator's problem is to devise chemical means of reproducing the activity of plant root systems. Often success is minimal. In the case of lettuce plants, the amount of lead found in the plant can be related to the amount of lead extracted from the surrounding soil by molar nitric acid. With oats, lead uptake correlates better with the amount of lead extracted by a milder treatment (e.g., 0.01 M nitric acid or molar ammonium acetate). The application of copper sulfate, either to soils or as a foliar spray, increases the copper content of wheat plants, but in this case the plant metal uptake does not appear to correlate with the amount of available copper as determined by extraction with either strong acid, or molar ammonium acetate, or strong complexing agent (EDTA).

When foliar sprays are involved, one would expect some variation due to direct sorption of metal ions by the exposed leaf surfaces. A similar mechanism appears to apply in the accumulation of lead by plant matter growing near motorways.

It is tempting to take the argument to an extreme and suggest that monitoring of soil or atmosphere levels has little value if one is concerned primarily with potential dangers associated with the incorporation of toxic materials into the food chain.

However, despite the obvious limitations, monitoring of single phases will continue, since such studies facilitate the identification of zones of contamination; permit recording of intermittent changes in levels; and supply guidance in respect to the role of external factors such as wind direction, rainfall, topography, etc.

For the recording of trends, several marker systems could prove to be equally appropriate, provided one does not attempt to place undue emphasis on the relative magnitude of the numbers obtained. For example, Table 1.2 lists some selected data from a pollution study in which up to four marker species were analysed. Each set of data

Table 1.2

Effect of Geographic Direction from Source on the
Concentration of Heavy Metal Ions in Samples[a]

Sample point	Nature of sample			
	Grass	Lichen	Moss	Soil
Pb, ppm				
A	10	130	120	—
B	49	1528	1200	—
C	86	—	—	270
D	160	—	—	230
Zn, ppm				
A	102	675	1213	—
B	146	1135	4870	—
C	350	—	—	450
D	270	—	—	416
Cd, ppm				
A	8	68	93	—
B	13	83	137	—
C	9	—	—	7.1
D	9	—	—	7.7

[a] Data from Burkitt, A., Lester, P., and Nickless, G.: Nature, 238: 327, 1972.

[b] All points approximately 2.6 m from smelter units.

confirms that the degree of pollution varies with geographic direction (sites A, B, C, and D) but if desired, individual numbers could be selected to support claims that one area is either twice, four times, or ten times as polluted as another.

Similar considerations apply to the monitoring of any water supply. Examination of filtered samples simplifies analysis but does not necessarily provide meaningful environmental data, since significant amounts of many water-quality constituents are transported in streams while attached to suspended sediment. In addition, settled sediment can act as a reservoir, releasing toxic materials into the food chain or aquatic life during times of chemical change (e.g., influx of water polluted with acids, organic matter) or times of significant turbulence.

Sediments are often complex mixtures and in oceanographic studies, selective chemical treatments are being used to distinguish between detritic and nondetritic components, with the latter being roughly subdivided into adsorbed species, hydrous oxides, and metals associated with organic matter. Within these groups, there are various interactions, e.g., colloidal clay particles in salt water can adsorb more than 2.5% of humic acid and the presence of the humic acid (derived from soil and plant life) modifies the adsorptive capacity of the suspended material. Changing the electrolyte strength reverses the process (in fresh water the amount of the adsorbed humic material is less than 0.4%) and, accordingly, at the points where fresh and salt waters mix, e.g., estuaries, there can be marked changes in the distribution of heavy metals between suspended material and the aqueous phase.

Since living marine organisms tend to accumulate high concentrations of heavy metals when exposed to polluted environments, it has been proposed that such materials should be used as monitors. The validity of this approach is challenged by the results obtained in surveys of the distribution of heavy metals in shell fish. As in the case of plants, the metal ion content of a given organism varies (by as much as a factor of ten) in accordance with its location relative to potential pollution sources, but it is equally apparent that within a given sampling region, there can be major differences in the determined values.

At a given site, variations in the apparent composition of selected markers can be of the order of $\pm 20\%$ of content. Accordingly, differences between sites need to be large in order to be statistically significant. In part, this lack of precision can be attributed to the fact that one is dealing with heterogeneous equilibrium systems.

C. Influence of Competing Equilibria

Let us consider the data reported in Table 1.2 in terms of the equilibria involved. (Shown schematically in Fig. 1.1.)

Since the moss used was suspended in nets from trees, any accumulation of metal ions can be attributed primarily to concentrations of particulates and vapors in the surrounding air. The material is reported to have some ion-exchange properties, hence losses due to solution (e.g., in rain) may have been reduced by exchange reactions.

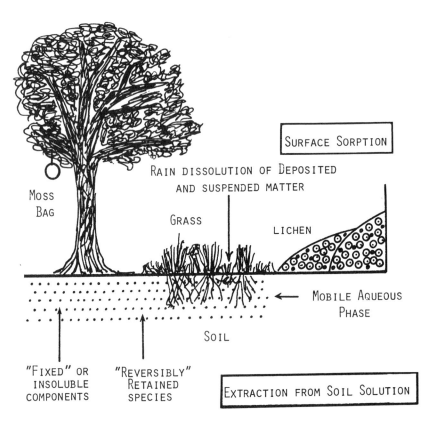

Figure 1.1   Diagram showing alternative modes of pollution uptake by a simple environmental system.   Total content found in any "marker" species can include contributions from surface adsorption, ion exchange, or root feeding with source material being conveyed by gaseous or liquid phases.

This secondary process is less likely with the ground-level collectant, lichen.

Grass is also a ground-level collectant but in this case, some fraction of the metal content could have been derived by plant uptake from the soil (the leaves of the tree, if studied, would also have had a dual source of contamination).   The extent of the uptake from the soil can be predicted to vary with factors such as the concentration of salts in the soil solution, the chemical nature of the metal species, the sorptive behaviour of the soil components, pH of the system, etc.   It is a

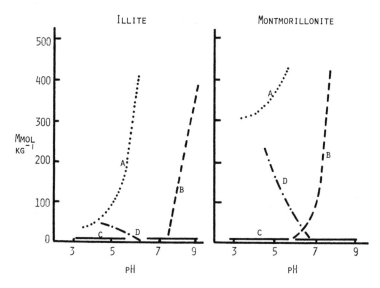

Figure 1.2 Adsorption of copper species from solution by clay suspensions (0.05% w/v) as a function of pH. Initial concentration of
copper in solution 10 mg/l. (a) No ligand present, precipitation of
hydroxide at pH > 6; (b) molar ratio of oxalate to copper present, 6:1
(note how onset of precipitate formation is displaced to higher pH by
oxalato complex formation); (c) equimolar ratio of EDTA and copper
ion-anionic complex stable over total pH range-no adsorption; (d)
molar ratio of cyanide to copper present, 6:1 (note effect of pH).

complex process only partially understood. The marked influence
some variables can exert may be gauged from Figure 1.2 which shows
the effect of pH and ligands on the sorption of copper ions by two common clay minerals, illite and montmorillonite.

   The extent of sorption, by either organic or inorganic substrates,
varies in magnitude in accordance with the affinity of the species for
the particular adsorbent (the sorption coefficient) and with the concentration of adsorbable material.

   The amount retained is reduced when there is competition for the
adsorption sites. If the system is not too complex, its behavior may
conform to a mathematical expression of the type:

$$\left(\frac{x}{m}\right)_A = \frac{k_1 C_A S_V}{\{1 + k_1 C_A + k_2 C_B + k_3 C_C + \cdots\}}$$

where $(x/m)_A$ is the amount of species A sorbed per gram of adsorbent; S_V is maximum or saturation value; C_A, C_B, C_C are the equilibrium concentrations of the competing species, A, B, and C respectively; and k_1, k_2, k_3 are the corresponding sorption coefficients. It can be observed that if k_2C_B or $k_3C_C \gg k_1C_A$, the value of $(x/m)_A$ is decreased markedly.

Many of the systems regularly subjected to trace analysis involve as many sets of competing equilibria as the example shown in Figure 1.3.

Prediction of the effect of changes (in say, pH) on this kind of system requires simultaneous consideration of sorption isotherms, ion-exchange relationships, solubility products, and stability constant equations.

Not specifically mentioned in this cycle is the important role that may be played by bacteria. Consider the case of mercury pollution. The transformation of inorganic mercury compounds to methyl mercury (II) ions (CH_3Hg^+) is believed to occur mainly on the surface layers of sediment or suspended organic particles. The conversion rate depends somewhat on the ratio of mercury input to organic loading and has been estimated (for a Swedish lake) at about 0.1% of the total mercury present per year. The organometallic form is readily assimilated by plants and fish, hence it tends to become concentrated in the biota. This leads to the following type of mercury distribution within a lake, or closed estuary, system: Sediment, 90 to 99% (about 10% as CH_3Hg^+); water phase 1 to 10% (almost entirely as inorganic forms bound to suspended particles); biota < 0.1% (predominantly CH_3Hg^+). Specific members of the marine biosphere (e.g., shellfish) tend to retain the metal species strongly and concentration factors as high as 10^6 have been found.

There are some mercury-resistant bacteria that bring about a reverse process, that is, organomercury pollutant species are converted into insoluble free mercury. In addition, under anaerobic conditions, sulfate-reducing bacteria can produce sufficient hydrogen sulfide to precipitate mercury(II) ions as the sparingly soluble sulfide; this compound is not transformed by bacteria into methyl mercury.

Consideration of the multiple equilibria associated with most natural systems not only reemphasizes the problem of marker selection but illustrates the difficulty associated with ensuring minimal changes during sampling and storage of samples. Theoretically, one needs to "freeze" all chemical and bacteriologic transformations, or devise suitable in situ procedures. Effects may be minimized by storage at

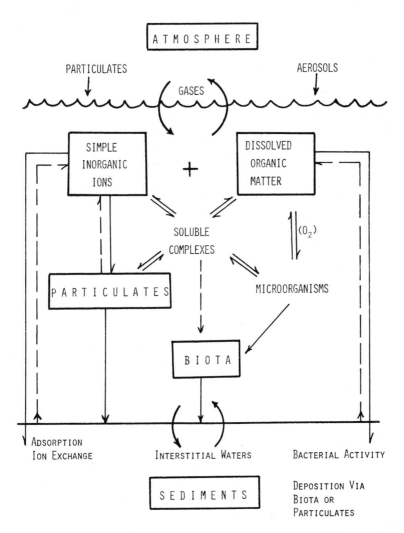

Figure 1.3 Diagram showing some of the competing equilibria and interactions which can be operative in a natural water system.

low temperatures, or by the use of mobile laboratories, but doubt must always remain since the simplest basic process may disturb the balance of the system. For example, removal of a water sample from above a sediment (or separation of the suspended matter) eliminates one component and allows solutes to establish new equilibrium positions,

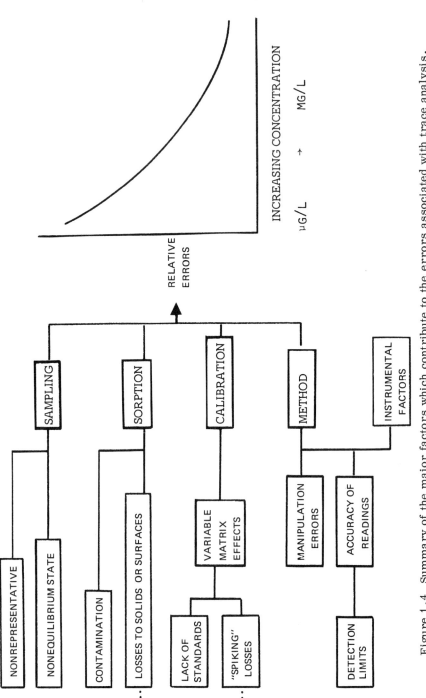

Figure 1.4   Summary of the major factors which contribute to the errors associated with trace analysis.

particularly with respect to extraneous sorption sites such as the walls of the retaining vessel. Introduction of a competing species (e.g., acidification) can solve this sorption problem, but what effect may this have on the distribution between other chemical forms? It can be argued that redistribution is unimportant where total contents are adequate markers for pollution; but even in such cases one must ensure that the analytic procedure measures the total, (that is, responds to all chemical forms). For example, testing a natural water by atomic absorption spectroscopy normally gives the sum of hydrated ion, complex ions, and colloidal components. An ion-selective electrode measurement yields values for the first of these only.

This returns us to the analytical problems associated with pollution evaluation. The method of analysis chosen must measure, accurately, the desired chemical component and possess a speed, simplicity, and sensitivity appropriate to making multiple determinations in reasonable time allocations. Allowance has to be made for interference effects, i.e., errors introduced by the presence of other species in the sample, and for artifacts, such as sorption losses or contamination on storage.

Collection of valid data thus requires both careful assessment of the most appropriate marker and minimization of the multiple sources of errors associated with trace analysis.

A summary of some of the factors that contribute to the total error is given in Figure 1.4, while Figure 2.1 indicates the type of variations that can be observed in interlaboratory comparison studies.

Reputable investigating groups recognize the various problems as outlined in previous sections and have devised compromise answers for most situations. They attempt to ensure that their data is adequate for assessment of the overall situation, and sufficiently reliable for regular comparisons. Deficiencies are acknowledged and form the subject of continuing research. Eventually this attitude will spread so widely that reliable defensible data, and not mere numbers, will be the norm.

2

SIGNIFICANCE OF SAMPLING AND STATISTICS

I. ACCURACY AND PRECISION

Since it is impossible to count every atom in a given sample, the exact composition of a material is never known, although it is possible to determine the most probable composition. The determination of this true result may require that the material be analyzed carefully many times by a variety of techniques. (Materials subjected to such multiple analyses are often retained for reference purposes and are then known as "standards.")

The nearness of an experimental result to the true answer is termed the accuracy of the procedure, while precision is a measure of the reliability or reproducibility of a technique.

Among the many sources of variation in experimental results are the following:

> Errors inherent in the method of sampling and method of analysis
>
> Errors introduced by the analyst due to personal bias or carelessness
>
> Errors introduced by variations in the quality of chemicals or performance of instruments

Two types of error can be distinguished: systematic and random. Systematic errors are those which tend to give results that are always higher or lower than the true figure. Random errors lead to results that are sometimes higher and sometimes lower than the true result. Accuracy measures random plus systematic errors, precision is a measure of the purely random errors.

The distinction between accuracy and precision may become more apparent from a study of Figure 2.1.

The thick horizontal lines on this diagram represent the amount of individual component added to a synthetic mixture (i.e., the "true" results). The concentrations actually reported by individual laboratories varied in magnitude over the range indicated by the limits of the various rectangles. Each laboratory achieved a satisfactory degree of precision (that is, replicate analyses varied little from an experimental mean), but only those who reported a mean value similar to the true result could claim to be accurate.

It should be noted that the mean of all submitted results (dotted lines) did not closely approximate the amount initially taken on all occasions.

The difference between an experimental mean and the true results (in any system), yields an absolute error value. For comparison purposes, the relative percentage error (i.e., 100 X difference/true result) is often more informative. For example, the relative percentage error associated with the mean values for Cr and Al (Figure 2.1) are similar (ca. 2%) but the absolute errors (expressed as mg/liter) vary by a factor of four.

Where a large number of determinations have been made on a particular universe, the results can be arranged in the form of a distribution curve.

A distribution curve records the number of samples (or determinations) whose composition has been found to lie within small restricted ranges. Thus in the example shown in Figure 2.2, of 100 samples analyzed, 5 yielded results lying between 10.0 and 10.2%, 10 results were obtained in the range 10.2 to 10.4%, 20 results fell between 10.4 and 10.6, etc. With a very large number of samples and a narrow range for each group (e.g., 0.01%) a smooth curve is obtained. In the example given in Figure 2.2, the curve is symmetrical in shape, and approximates what is known as a "normal" (or Gaussian) distribution. Not all analytic distribution curves are normal, and the discussion which follows is only strictly applicable to those systems whose behavior approximates this ideal. [For example, systems in which the variance is small (e.g., less than 5 to 10% of the total concentration)].

The spread of distribution reflects the magnitude of random errors and thus the precision of a process. In the example shown in Figure 2.2, 70% of the results lie in the range 10.4 to 11.0%, while 90% of the results lie in the range 10.2 to 11.2% More precise results would

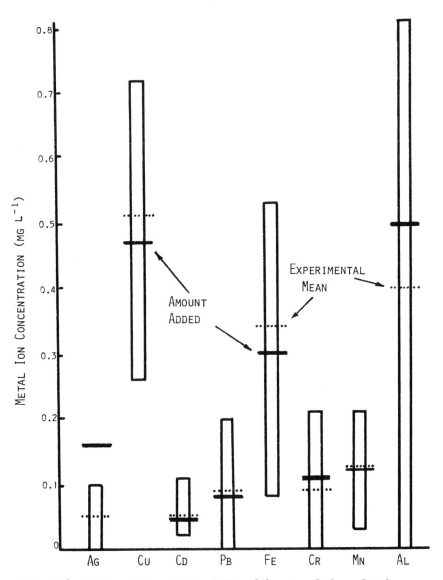

Fig. 2.1 Diagrammatic representation of the spread of results ob-
served in an interlaboratory comparison study. The concentrations of
trace metal ion were determined by standard colorimetric procedures.
(Based on data reported by Hume, D. N.: "Progress in Analytical
Chemistry," Vol. 5, "Chemical Analysis of the Environment," Plenum
Press, N.Y., 1973, pp. 3-16.)

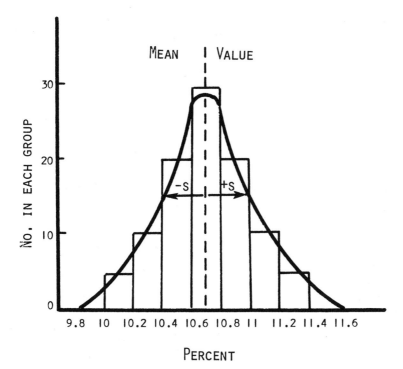

Fig. 2.2 Histogram illustrating the random distribution of a series of analytic results.

give a narrower spread of analytical readings. If it is necessary to express precision as a number, then the value reported has to be some measure of the observed spread of results.

With a random distribution, the best probable value of the sample composition is the <u>mean</u> of all the values obtained, and various methods have been proposed for recording the spread of results in terms of deviations from this mean value. In one method, the mean deviation is calculated, but this approach has been replaced almost completely by the use of standard deviations:

Mean deviation $= \dfrac{\Sigma d}{n}$

Mean square deviator or variance $V = \dfrac{\Sigma d^2}{n}$

Standard deviation $= V^{\frac{1}{2}} = \left(\dfrac{\Sigma d^2}{n}\right)^{\frac{1}{2}} = \left[\dfrac{\Sigma (x - \overline{x})^2}{(n - 1)}\right]^{\frac{1}{2}}$

$\quad$ (σ or S)

where d is the deviation of the results from the mean value (independent of sign), Σ d represents the sum of all d terms, i.e., $d_1 + d_2 + \ldots + d_n$, and n is the number of samples examined. Where n is small, e.g., less than 10, it is statistically preferable to divide the deviation terms by (n - 1). The symbol σ is used if the results refer to the standard deviation of a large population, while S refers to the deviation of samples taken from the larger population. S may be regarded as an estimate of σ, and x represents an individual result, while $\bar{x}$ represents the mean of all n results.

The values $+\sigma$ and $-\sigma$ indicate the points of inflection of the Gaussian curve of error, and the probability of one result falling within $\pm \sigma$ of the mean $\bar{x}$ is approximately two out of three.

Since it is possible for one in every three results to be outside the range of $\bar{x} \pm \sigma$, many consider it preferable to state precision in terms of confidence limits. In this terminology, one expects a given percentage of the results to lie within stated limits; the percentage commonly taken being 95. For normal distributions, the 95% limits are $\bar{x} \pm 1.96\sigma$.

An alternative method of treating the results is to consider the odds against obtaining a value outside certain set limits. The quantity P is defined as the probability that the limits calculated will be exceeded by chance. For a given probability, the limits for a given set of data are given by $\bar{x} \pm \sigma t$, where t is a factor whose magnitude varies with the value assigned to P and with the number of degrees of freedom (n - 1). Tables are available which list the values of t for varying values of P and (n - 1).

For example, if results are initially collected from 31 tests and P is limited to 0.01, one out of every hundred subsequent answers would be expected, on the average, to be outside the limits of $\bar{x} \pm 2.750\sigma$; t being 2.750 for (n - 1) = 30 and P = 0.01.

If the theoretical or "true" value of a sample is known, a value for t can be calculated from the equation: $t =$ (true value - experimental results)$/\sigma$; and the probability tables can then be used to test the significance of results.

In this case, one has a calculated value for t and a known number of degrees of freedom (n - 1); reference to the tables gives a probability value P. It is sufficient in most cases to know that the value is less than 0.05. When the probability found by such a test is less than 0.05, it indicates that there is less than a 1 in 20 risk that an incorrect decision has been made with regard to the validity of the result or with regard to the truth of the statements used to obtain the result. The result is therefore significant.

This definition of significance is a mathematical convention which should be used on results derived from real systems only if its frailties are fully appreciated. For example, if a substance or property of a system is subject to daily monitoring, then more than one sample a month may show a value significantly different from the true value, by chance.

When the experimental results being considered are the mean values derived in several (n) series of tests, the standard deviation term used for calculating t is taken as $\sigma/(n)^{\frac{1}{2}}$; this is known as the standard error of the mean, $\sigma_{\bar{x}}$.

The accuracy of an analytic method can also be described in terms of a standard deviation. In such calculations, deviations from the true value are substituted in the basic equation instead of deviations from the arithmetic mean of the series of results.

Decisions on whether an individual result can be legitimately rejected from a series can be based on the error function σ. Often, results outside the 2σ or 3σ limits from the mean of the total number of observations are rejected. In special cases, the 4σ limits are chosen as the arbitrary points for rejection of results.

The results produced in analytical laboratories form the basis for many types of decisions, and if correct decisions are to be made, every care must be taken to ensure that errors in the analytical results are minimal. Simple statistics, as outlined above, provide a means of measuring the magnitude of the inherent errors. The calculations can also be used to evaluate the effect of variations in the analytic procedures. The statement that a particular result has a value of $\bar{x} \pm \sigma t$ gives an indication of the reliability of the data to the decision maker; it should also encourage the analyst to attempt to reduce the numerical value of σ in subsequent analyses.

Many factors contribute to the magnitude of a standard deviation, but improvement can usually be achieved by greater care in manipulation, by the study of a larger number of samples, and by slight modifications in procedure.

It should be emphasised that discrete values of quality characteristics are not always distributed in accordance with a pure Gaussian curve. In fact, much of the data derived from environmental studies yields skewed curves and more closely conform to a log-normal distribution. That is, the variable x has to be transformed into log x in order to approximate the shape of a normal curve.

Probability-paper plots can be used to identify the distribution type. On this special paper, the probability axis records the percentage of

total values which fall below the experimental value indicated by the
other axis. With arithmetic probability paper, the observation scale
is linear, and this form is used for normality testing; log probability
paper has a logarithmic observation scale and is used to test for log
normality. The more closely the graph approximates a straight line,
the better is the evidence for the adequacy of the chosen model.

An example of each type of distribution is shown in Figure 2.3.

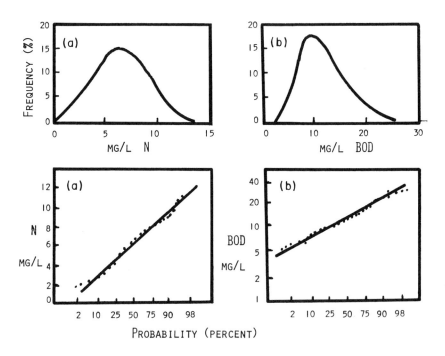

Fig. 2.3 Distribution curves and probability plots, of best fit, for
two sets of analytic data collected in a river system over a period of
time: (a) Ammoniacal nitrogen content: distribution normal; (b)
biologic oxygen demand; log-normal distribution. (Based on data
reported in Department of Environment Leaflet No. 54, H.M. Stationary Office, London, 1971.)

Before concluding this section on accuracy and precision, some brief comments should be made on the use of significant figures.

In general, the results of experimental observations should be reported in such a way that the last digit given is the only one whose value may be in doubt. These digits (excluding leading zeros) are termed significant figures.

The uncertainty level of data may be derived from standard deviation calculations, or estimated from known limitations in the measuring technique.

As an example, let us consider an assay in which 1-g samples of material are to be weighed on an analytic balance, and after dissolution, are to be titrated using a 50 ml burette. If the precision of the balance is ± 0.1 mg, uncertainty in the sample mass occurs only at this level, and the mass (in g) should be reported to four decimal places (e.g., 1.0012). The burette barrel is marked in 0.1 ml divisions and uncertainty is introduced when one tries to interpolate between divisions. Accordingly, individual titration volumes should be read to a second decimal place (e.g., 11.65 ml). However, the standard deviation associated with a series of such titrations may be found to be, say ± 0.21. This deviation value indicates that in the overall process, there is uncertainty in regard to the first decimal place, and the reported mean should be rounded off to this level (e.g., 11.7 not 11.67 ml).

Before final recording, most experimental data has to be subjected to some form of mathematical manipulation, and it should be remembered that the final answer cannot be more accurate than the least accurate datum involved.

Consider a computed result R derived from the measured quantities A, B, and C. Let us assume that with each measurement there is a known determinate error (e.g., half the smallest division on the measurement scale) and let these errors be represented by α, β, and γ respectively.

If the mathematical operation required is addition or subtraction (e.g., $R = A + C - B$), then it can be shown that the resulting error in $R(\delta)$ is given by $\delta = \alpha + \beta + \gamma$.

On the other hand, where multiplication and division are involved (e.g., $R = A \cdot B/C$), the values transmitted into the corresponding equation are the relative determinate errors, i.e., $\delta/R = \alpha/A + \beta/B + \gamma/C$.

In chemical determinations, indeterminate errors can be of greater concern. Such errors are manifested by a scatter in the data when a measurement is performed more than once. The degree of scatter is indicated by the standard deviation, and it is the square of this value (known as the variance) which is useful in estimating propagated indeterminate errors.

In addition or subtraction operations, the variance of the result equals the sum of the component variances, i.e., if

$$R = A + C - B, \quad s_R^2 = s_A^2 + s_B^2 + s_C^2$$

With multiplication or division of experimental results, it is the squares of the relative standard divisions which are transmitted, i.e., if

$$R = A \cdot \frac{B}{C}, \quad \left(s\frac{R}{R}\right)^2 = \left(s\frac{A}{A}\right)^2 + \left(s\frac{B}{B}\right)^2 + \left(s\frac{C}{C}\right)^2$$

In many analytic calculations, only one term possesses a significant relative error (e.g., the titration volume in the preceding example), and in such cases the percentage relative error of the calculated result is similar to that calculated for the major contributor. This relationship must be reflected in the number of significant figures used in the final reported result. For example, a reading of 11.7 implies an uncertainty (or possible error) of ± 0.1, or a percentage relative error of approximately $\pm 1\%$. If this reading is multiplied by say, 0.0957 and divided by, say 1.0012, on a calculator, one obtains an answer having up to 10 digits; but to meet the percent relative error restriction the result has to be rounded to 1.12 (i.e., back to three significant figures).

Failure to indicate the probable error of a laboratory result (e.g., through significant figures or quoted standard deviations) can lead to doubtful interpretations of effects. In studies of natural systems this aspect is compounded by the variance of sampling. In many situations the latter variance is the dominant source of propagated indeterminate error, and if taken into consideration, one may need to reduce the number of significant figures used. For example, if the laboratory result of 1.12 noted above was derived from an environmental sample, poorly collected and stored, a more realistic value for final reporting could be 1.1 or 1.

II. GENERAL CONSIDERATIONS IN SAMPLING

A. Collection of Gross Sample

Since small sample weights or volumes are used in laboratory studies, it is essential that the material examined should be truly representative of the whole universe of interest. In the ideal situation, the result obtained from the small laboratory sample should be identical with that obtained by collectively examining the whole mass of material.

The art or science of obtaining small amounts of material which approach this ideal is known as sampling.

As an example of the importance of sampling, consider the data shown in Table 2.1, and note the variations in the degree of salt contamination of sand taken from different areas of a small beach.

Unless a bulk mass of material has been sampled very carefully, the results obtained in a most careful chemical analysis may prove to be absolutely useless. It must always be remembered that the overall error of an analytical result is composed of both sampling errors and errors inherent in the analytic technique (Fig. 1.4)

If one assumes that the experimental mean of a series of studies closely approximates the true result, the overall accuracy A of the total procedure is described by a relationship such as

$$A = 2(V_s + V_p + V_a)^{\frac{1}{2}}$$

where V_s = variance of sampling; V_p = variance of preparation; and V_a = variance of analysis.

Table 2.1

Salt Contamination of Beach Sand (% NaCl)

Sample site location	Particle size[a]		
	+18 Mesh	−18 + 85	−85 Mesh
Far inshore	0.001	0.001	0.001
Beach center	0.01	0.006	0.01
Near waterline	0.01	0.2	2.2

[a] Particle-size distribution varied with location of site but approximated 3% + 18; 8% + 30; 34% + 44; 45% + 60; 9% + 85; and 1% − 85 mesh.

The initial step in sampling is generally the collection of a <u>gross sample</u>. To obtain a gross sample, a large number of increments of the material (henceforth called the "universe") are gathered and mixed together. The individual increments should all be of the same size and should be collected in some manner which eliminates any chance of personal selection. For unbiassed sampling, all parts of the universe should have a chance of being sampled, and the best way to eliminate bias is to use a procedure of random sampling. However, for chemical and physical tests, the increments are usually collected systematically from areas spaced evenly over the whole universe.

The method by which a sample is collected is determined by the nature of the material being handled; and the minimum satisfactory weight for the gross sample depends on the sample type (e.g., grain, crushed solids, metal, liquid, gas, etc.), and the quality or homogeneity of the material. The amount of gross sample required tends to decrease with decreasing particle size (i.e., of crushed solids) and increase with increasing degrees of heterogeneity.

The size of each increment should be large enough to avoid bias with respect to particles of different size or nature. The number of increments to be taken from the universe can then be adjusted to yield some desired accuracy level in the final analysis. One formula for calculating the number of increments N required is $N = 4V/A^2$ where A is the accuracy figure arbitrarily selected by the analyst and V is an estimate of the overall variance (i.e., $V = V_s + V_p + V_a$). This estimate (V) is derived through analysis of a series of samples taken from the bulk mass in a preliminary study.

Since the true result is rarely known, it is more realistic to define, at some stated level of confidence, the precision p sought. Using these terms, the number of samples required is $(k\sigma/p)^2$, where k is a coefficient whose magnitude depends on the confidence level chosen. Some typical k values are given below:

Confidence level (%)	99	98	95	90	80	68	50
k	2.58	2.33	1.96	1.64	1.28	1.00	0.67

Where one is dealing with desired or permitted tolerance levels it can be more informative to calculate the <u>percentile values</u> (Pv), that is, the value exceeded on (100 - Pv) percent of occasions. In order to estimate a percentile value, with a selected degree of confidence, it is necessary to collect $(fk\sigma/p)^2$ samples, i.e., f^2 times the number required to estimate a mean with the same degree of confidence. This increase arises because, in a collection of data, more values fall close to the mean than close to any selected value away from the mean.

Values of f are as given below:

Percentile	50 (median)	40	30	20	16	10	5	1
		60	70	80	84	90	95	99
f	1.25	1.27	1.32	1.43	1.52	1.71	2.09	3.67

Any calculated value of N is itself an estimate since its validity depends on the accuracy with which the initial V ($\equiv \sigma^2$) predicts the true distribution of the results derivable from the bulk mass.

If the variable x in a system is distributed log normally, the transformed variable log x is distributed normally; and with this modification one can handle log-normal distributions in a similar manner (mathematically) to the cases discussed above.

To establish any sampling program on a sound scientific basis, it is essential to understand the relationship between final precision and the number of samples required to obtain it. The question that the information to be gathered is intended to answer must be clearly defined, so that one can ensure that the precision selected is adequate for the problem. At the same time, seeking greater precision than actually needed can be expensive in terms of sampling time, because high precisions require the collection of very large numbers of samples.

B. Reduction of Sample Size

Except in those cases where collection of the gross sample involves a prolonged time sequence, individual increments are usually mixed together to form a bulk sample. Such bulk samples are generally too large for direct transportation to a laboratory, and some subsampling is normally undertaken.

With gases and liquids, subsamples are taken after an efficient mixing stage. In this process, one must take care to minimize side-reactions, since these can invalidate results.

With particulate matter, it is necessary to reduce the size of the particles before mixing and resampling. This is a corollary of the recommendation that larger gross samples are required for material of larger particle size. For example, in the sampling procedure for one particular mineral, it is suggested that no less than 500 lb of gross sample should be collected if the diameter of the largest lump is 1 in., while a gross sample of 4 lb suffices if the diameter of the same material is 0.13 in. This ratio of weight of sample to particle size is known as the size-weight ratio and in preparation of the laboratory sample this ratio must always be considered. Thus if one collects 500 lb of 1-in. lumps and wishes to transport only 4 lb to the laboratory, all of the 500-lb sample should be crushed to a particle size of less than 0.13 in. The crushed material should then be intimately mixed, and a 4-lb sample obtained by collecting a number of increments from over the whole new universe of crushed mineral. The process of size reduction, mixing, and sampling may have to be repeated several times before the final collection of the finely ground laboratory sample.

As shown by the equation for overall variance quoted previously, errors introduced in the preparation stage are as important as errors introduced in initial sampling or final analysis, and hence due care is required in all preparation procedures.

The most suitable method for sampling any particular universe is difficult to specify, since each case has to be considered on its merits. Ideally, each sampling (and subsampling) procedure should be examined statistically to elucidate the conditions required to achieve desired precision limits. This tends to be a prolonged operation, and most workers are content to adopt the recommendations of responsible bodies such as standards associations (e.g., American Society for Testing Materials, British Standards Association, American Public Health Association, and environmental protection agencies).

In the sections which follow, some of the general considerations which apply to a few classes of universes are considered, and this should serve to indicate the flexibility of mind required when planning sampling programs for a range of materials.

III. VARIABLES ASSOCIATED WITH NATURE OF PHASE

A. Sampling of Liquid Systems

Where a liquid system can be demonstrated to be essentially homogeneous, a single small-volume sample can often adequately represent the composition of the bulk universe at a given time. For example, with turbulent streams, or streams of limited cross section, collection of samples at middepth in midstream is usually satisfactory. Similarly, one sampling point is usually sufficient when testing reservoirs and storage tanks, but the absence of stratification should be confirmed by initially taking a series of samples at different depths. Should sampling involve drawing the liquid through pumps, pipes, or taps, several gallons of liquid should be allowed to flow to waste prior to sample collection.

The quality of liquid samples can vary with time, and in such circumstances samples should be taken at regular time intervals. Good examples of time variant systems are effluent flows and rivers. The quality of both depends in a highly complex fashion on the vagaries of the weather and a wide range of human activities. Some are continuously variable, some are intermittent, and some even unique. Because of the continual variations, the ideal basis for making assessments is a continuous record, but currently only a few characteristics can be monitored in this fashion. It is thus necessary to rely on intermittent sampling and subsequent analysis. The discrete values of concentration or load (mass per unit time) often prove to be distributed according to statistical patterns with well-established properties (e.g., a normal curve), but such frequency distributions do not describe the sequence in which the fluctuations actually occur. The range of such variations may be very large, even over short periods; diurnal fluctuations of an order of magnitude, for example, have been observed. Multiple sampling points are needed in surveys of complex waterways, e.g., in the case of a river system one should preferably collect samples from all the important effluent outlets and tributaries just before they reach the main river. In the main river sampling should be undertaken from positions just above and below confluences with important tributaries and effluent outlets, below weirs and also above them if dissolved oxygen is being measured, and, if necessary, at intermediate points. The sampling points should be located well away from any possible disturbing influences, such as whirlpools, stagnant zones, heavy growths of weed or sewage fungus, or points where groundwater enters (unless it is desired to study specifically their effects on water quality).

Small differences in density (e.g., 0.001) or temperature (one or two degrees) can be sufficient to maintain persistent vertical stratification in rivers. Unless a zone of turbulence intervenes, gross lateral, as well as longitudinal and vertical differences in concentration often exist for many miles below confluence or discharge points. Even in estuaries where there is tidal flow, complex stratification patterns persist, and inputs of low-density waters can tend to flow over denser layers with minimal intermixing.

Unless the measurement of concentration is the only object of the program, provision should be made for measuring (or estimating) the liquid flow rate at all sampling locations at the time of sampling. Knowledge of the time of travel between sampling points can aid the interpretation of results and can help decide the best times for sampling different locations in the same system. (Tracer studies are useful for this purpose.)

Liquids moving along pipelines can be monitored by fitting a short bypass to which is attached an automatic sampling valve designed to take samples at predetermined time periods. This technique is particularly convenient if analysis is based on an instrumental technique, since small samples can be fed directly into the instrument at regular intervals without any manual operations.

The minimum volume of liquid required as a sample is determined by the number of tests to be made, and by the amount required for each test; usually a liter or two suffices. It is recommended that the samples be collected in carefully cleaned bottles made of pyrex glass, polyethylene, or other inert material. The choice of material and cleaning procedure depends somewhat on the nature of the determinations required. For some studies cleaning with detergent is satisfactory, for other determinations the initial wash should be followed by acid cleaning (e.g., chromic acid, water rinse, 8 M nitric acid). Glass containers for microbiologic samples have to be sterilized in an autoclave. For oil and grease or pesticide studies it is desirable to prewash the glass bottles with organic solvents, e.g., n-hexane and acetone. All sample bottles should be rinsed two or three times with the liquid to be collected before filling with the final sample.

On mixing of several aliquot samples, or on standing of a single sample, chemical or biologic changes can occur. For example, metal ions may precipitate as hydroxides or form complexes with other constituents; cations or anions may change valence states; other constituents may dissolve or volatilize with the passage of time. A good example of the latter is the loss of mercury through polyethylene (Fig. 2.4). Metal cations (e.g., iron and lead) can also adsorb onto the walls of the containing vessel.

These effects can be minimized by using a preservation procedure, such as pH control, chemical additions, or refrigeration to $4^{\circ}C$. Even with these precautions, whereever possible, the sampling program should be designed to provide the shortest possible interval between sample collection and analysis. Table 2.2 summarizes one series of recommendations with respect to suitable sample containers, preservation methods, and holding times.

In waterway surveys, it is useful to distinguish between the total content and the dissolved concentration of a species. The latter is defined as the amount present after filtration through a 0.45-μm membrane filter, and this filtration (if required) should be carried out as soon as possible after collection of the sample.

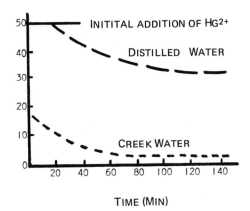

Fig. 2.4 Diagram illustrating the rate of loss of inorganic mercury
from water stored in polyethylene bottles. (Based on data from Coyne,
R. V., and Collins, J. A.: Anal. Chem. 44:1093, 1972.)

B. Gas Sampling

The techniques used for gas sampling may be divided into two major
categories, but whatever the technique, one has to remember that gases
mix readily, and it is easy to contaminate, or dilute, test samples with
"pure" air accidentally introduced in the sampling process.

    In the wet sampling approach, a clean glass vessel [e.g., a bottle
or gas sampling tube (compare Fig. 2.5)] is filled with a liquid which
has little solvent power for all the components of the gas mixture being
sampled (for many simple mixtures, acidified water is reasonably satisfactory). The liquid is subsequently drained out of the vessel in a
manner which allows it to be replaced by the atmosphere being sampled.
In the laboratory, the gas sample can be displaced and transferred to
the measuring apparatus by introducing the same liquid back into the
sampling vessel. Undesirable aspects of this approach include solubility
loss of some components in the residual films of liquid, introduction of
liquid vapors which may have to be removed before testing, and the in-
convenience of transporting and releasing large volumes of liquid.

    Dry sampling methods are thus often considered to be the preferable
alternative, but it should be emphasised that many samples have a natural liquid vapor content (e.g., water vapor) and condensation can yield
a liquid film capable of selectively dissolving some sample components
(e.g., SO_2, CO_2).

Table 2.2

Recommended Sample Containers, Preservation Methods and Holding Times for some Water Quality Parameters

Parameter	Container	Preservation	Acceptable holding time
Chlorine	Glass or polythene		None
Dissolved oxygen	Glass		
Hydrogen sulfide			
pH			
Sulfite			
Bacteriology	Sterile sampling vial	Sodium thiosulfate if Cl_2 present	6–12 hr
BOD	Glass or polythene	Refrigeration, 4°C	
Mercury	Glass	To pH 3 with HNO_3	
Nitrogen–Kjeldahl	Polythene	40 mg $HgCl_2$/liter; refrigeration, 4°C	
Acidity–alkalinity	Polythene	Refrigeration, 4°C	
Color	Glass	Refrigeration, 4°C	24 hr
Cyanide	Polythene	NaOH to pH 10, 4°C	
Oil and grease	Glass	2 ml H_2SO_4/liter; refrigeration, 4°C	

Phenolics	Glass	1 g $CuSO_4$/liter + H_3PO_4 to pH 4	7 days
Calcium	Polythene	5 ml HNO_3/liter	
Carbon, organic	Polythene	2 ml H_2SO_4 to pH 2	
COD	Polythene	2 ml H_2SO_4	
Fluoride	Polythene	None required	
Hardness	Polythene	None required	
Nitrogen: ammonia nitrate, nitrite	Polythene	40 mg $HgCl_2$1; refrigeration, 4°C	
Phosphorus	Polythene	40 mg $HgCl_2$/liter; refrigeration, 4°C	
Specific conductance	Polythene	None required	
Sulfate	Polythene	Refrigeration, 4°C	
Sulfide	Polythene	2 ml Zn acetate/liter, 4°C	
Total metal	Glass	5 ml HNO_3/liter	6 months
Dissolved metals	Glass	3 ml 1:1 HNO_3 to filtrate	

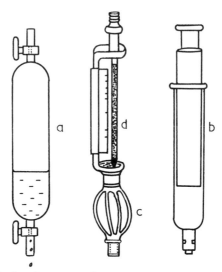

Fig. 2.5 Methods of gas sampling: (a) Liquid displacement using gas
sampling tubes; (b) syringe (or any other type of gas tight pump); (c)
rubber aspirator bulb, may be attached to a sampling line for filling
balloons or attached directly to measuring unit (as shown) (d).

In the dry sampling approach, gas is drawn through a perfectly dry
vessel by some form of pump or aspirator until the volume of the vessel
has been displaced at least six times. If piping or tubing is required to
connect the sampling area to the sampling vessel, it too must be flushed
thoroughly with the test gas. The flushing process can be avoided if
the sample vessel is evacuated prior to exposure to the test atmosphere,
but the slightest preliminary leak in the evacuated vessel leads to a major
sampling error.

In the laboratory, the gas samples are transferred into the measuring apparatus by liquid displacement (preferably mercury).

Modifications of the dry sampling approach eliminate the need for
subsequent liquid displacement.

In routine testing for minor specific components (e.g., alcohol
vapor in breath, toxic vapors in industrial air), the gas sample may be
drawn directly into a measuring device at the sampling site. The nature of the measuring device can vary from simple units such as adsorption tubes which selectively react with particular components to give

colored zones of size proportional to concentration, to expensive instruments such as the CO and SO_2 monitors described in the next chapter
(Section II).

For in situ collection of small volumes of gas, glass- or plastic-
bodied syringes can be extremely useful. Flushing and delivery to
testing equipment is facilitated by the moveable plunger, and attachment
of a needle allows samples to be collected from enclosed areas, e.g.,
pipes containing a sampling point sealed with a septum cap.

Another variation utilizes expandable collection vessels which can
be pumped up to pressures greater than atmospheric (e.g., balloons,
football bladders). The additional pressure aids the transfer of the
contents to measuring devices but at the same time creates a pressure
differential which promotes diffusion through the containing vessel
walls. Since the individual components of gaseous mixtures diffuse
through the polymer at varying rates, the samples should not be kept
in the bladders for many hours. On the other hand, if a series of
samples are to be collected from awkward positions or over a wide area,
it is often more convenient to collect and carry the samples in football
bladders and subsequently transfer the contents to glass vessels at a
chosen base.

Where specific components are highly soluble in water, they may
be isolated and preconcentrated by drawing large volumes of the gas
sample through an absorption unit which contains an appropriate sorbent
solution (i.e., water plus additives to enhance retention). The aqueous
solutions are subsequently examined by standard analytical procedures.

Drawing large volumes of gas through a filter pad of known porosity
allows collection of suspended particulate matter in quantities sufficient
for examination by microscopic and chemical methods.

If the gaseous material is already under pressure in the system of
interest, the introduction of gas sampling valves in various parts of the
process stream facilitates direct transfer of subsamples into analytical
apparatus or the filling of appropriate sample vessels.

Where gas flows are being examined (e.g., in smoke stacks), it is
desirable to draw samples from different parts of the flow and to collect the gross sample over a period of time (the logic of river sampling).
Simultaneous reading of temperature, pressure, and flow rates can aid
the interpretation of results.

C. Approaches to Sampling of Solids

The nature of the universe to be examined determines the approach most likely to conveniently yield a representative sample.

The sampling of a mineral deposit, ore body, massive rock, or soil stratum can be achieved by taking core drill samples at selected points over the whole surface area of the massive structure, the depth of drilling preferably covering the full thickness of the region of interest. By examining segments of the cores, the distribution of content as a function of depth can be estimated, and combination of the core results gives an average figure for the universe. In some situations examination of the exposed surfaces is considered sufficient. In this case fixed volumes of the solid are removed from regions of the exposed surface, the regions for sampling being selected either at random or in some systematic manner.

If the solid of interest exists as a pile of broken particles, increments have to be taken over the whole surface of the heap. The basic plan used to select the regions for increment removal must make allowance for the fact that in a mass of ungraded material, the composition of the dust is very probably different to the composition of the lumps. In a heterogeneous mass, the softer material may be different in composition from the hard lumps. Thus for shallow heaps, of limited dimensions, a suitable sampling method consists of complete removal of channels of material from across the whole width of the heap.

If the material to be sampled is being loaded or unloaded, it is sufficient to collect as the gross sample every tenth or twentieth shovel load or grab load. For bagged material, bags may be put aside at selected numerical intervals. Either the total contents or core samples from these bags can then be combined to yield the gross sample. With material flowing in a solid stream (e.g., on a conveyor belt) increments can be taken at regular intervals of time, each increment being drawn from across the full stream of material.

After collection, the gross sample should be passed through some form of mechanical device (such as a jaw crusher, a ball mill, or grinder) to reduce the average size of the particles. The ground material must then be thoroughly mixed before being divided into smaller heaps by either a set of riffles or by the technique of coning and quartering.

When the sample has been reduced to the required minimum particle size and the laboratory sample has been collected, it is common practice to dry the material before bottling in an airtight jar, but one should ensure that no component (e.g., sulfide minerals) becomes oxidized during the heating process. Occasionally, final grinding in a mortar and pestle may be necessary before chemical analysis.

With metal objects, the article is drilled or machined in a manner which allows material to be collected over a uniform cross section. The drillings or chips must be free of scale, surface metal, grease, dirt, or other foreign matter, hence neither water, oil, not any other lubricant can be used with the cutting machinery.

When molten alloys solidify in molds, segregation of components can occur, leading to zones of varying composition. Accordingly, the sampling area must be carefully chosen if it is to be representative of the whole. With nonferrous alloys, where segregation can be appreciable, a template is often prepared so that sample bars can be drilled in a regular sequential manner.

With plant material, the distributions of components can be very diverse and vary with age, hence a representative sample must contain material covering a wide spectrum of location and stage of growth. After drying, the organic materials can be crushed and subsequently mixed or blended to provide a final sample.

On standing, solid samples can deteriorate (e.g., rusting of steel drillings, loss of volatile components, etc.) but of more importance are the errors introduced by dissolution processes. Incorrect choice of acid can lead to partial solution of some components or loss of volatile products. With organic-type samples, a poor ashing technique can leave much of the species of interest in a form not detected by the selected analytical procedures. These aspects of sample treatment can enhance the variance of preparation (V_p) to such a degree that it becomes a limiting factor.

D. Definition of Sampling Objectives

Considerable attention is currently being paid to defining the objectives of analysis in relation to the use which will be made of the results.

As noted previously, it is equally important to first define the objective of the sampling program.

For commercial materials the primary aim is generally to establish the quality of a raw material source or to certify that a product has acceptable specifications. The precision tolerated depends greatly on the relative cost of the raw material or the stringency of the specifications.

With routine operations, the analytical data produced using samples collected in some systematic way, can be statistically processed at regular intervals in order to provide regular estimates of the mean and the total variance. Using this refined data, the sampling program may be modified to ensure acceptable precision and confidence limits.

With nonroutine or highly variable systems (e.g., pollution testing), the initial studies are likely to be very limited in precision due either to the small number of samples studied or because the aim is to establish the effect of location, wind direction, current flows, etc.

Once suitable sampling points have been selected, it is next necessary to decide if the aspect of major interest is peak concentration or some form of time-averaged mean. Detection of peak concentrations normally requires continuous sampling and recording, and, as such, tends to be more expensive. With time averaging, the big variable is the time allowed between sample collection; if too long, major changes in composition can go completely unnoticed.

The efficiency of a sampling procedure can only be assessed by regular testing and prolonged study of the background variability. Little confidence can be placed in the data derived from odd samples collected intermittently at random points.

EVALUATION OF ATMOSPHERIC POLLUTION

Since human beings breathe in large volumes of air each day in order to live, any contamination of the atmosphere is of utmost importance. However, assessment of air quality is difficult because the nature of atmospheric pollution ranges from visual restrictions caused by suspended particles to health hazards induced by trace amounts of toxic gases and vapors, and the extent of pollution varies with time and weather conditions. One should, therefore, sample over extended time periods and make suitable corrections for meteorologic variables, such as wind velocity, wind direction, temperature inversions, and so on. Some gases are water soluble and are lost in condensed vapor during sampling, others (e.g., oxides of sulfur) are very prone to sorption on fine solid particles.

The techniques used to monitor air quality vary in form from very crude to highly sophisticated, and with many the amount of pollutant present normally falls close to the detection limit of the technique. As a result, analytical precision tends to be poor (e.g., $\pm$ 20%), and adequate calibration is a challenge.

The objective, in this chapter, is to outline some of the procedures which have been used by various investigating groups. The emphasis is on principles, the possible significance of data, and approximate concentration ranges involved.

For detailed procedures, and information on biological effects, the reader is advised to consult specialist monographs such as those listed at the end of the chapter.

I. PARTICULATE SUSPENSIONS

A. Haze and Total Fallout

Suspensions of fine particles in the surrounding air obscure light and are visually objectionable. On settling, they lead to soiling (e.g. , soot on clothes), discoloration (e.g., of buildings), and sometimes corrosion (if particles are of active chemical nature).

Particle size distributions vary, but normally over 90% of suspended material has a diameter greater than 30 μm, with a maximum value of about 1000 μm. The larger particles rapidly settle out under the influence of gravity, but the smaller or microscopic components (i.e., < 100 μm) can be maintained as a suspension by eddies and air currents for indefinite periods.

Particles larger than 5 to 7 μm rarely enter the lungs, and only particles 0.25 to 5μm in diameter are believed to be effectively retained in the lungs.

Haze density readings, as measured by a nephelometer unit (Figure 3.1), provide comparative information on the effect of suspensions on light transmission. Air is drawn at a fixed rate (e.g., 0.01 m^3/min) through an optical maze to a chamber illuminated by a light beam. Some of the light is scattered by the fine (< 20 μm) particles present, and the intensity of this scattered radiation is measured by a photomultiplier tube mounted at right angles to the primary light beam.

The instrument response only accurately reflects changes in dust concentrations if the suspensions examined are similar with respect to particle nature and size distributions. Calibration in absolute units is very difficult.

In another type of unit, a fixed volume of air is drawn through a white filter medium, with subsequent study of the "darkness" of the stain. This provides comparative information on the amount of colored suspended matter in the atmosphere.

For identification of species, determination of size distributions, and provision of data on the amount of fine material present (i.e., that most closely related to health problems), a total sampling technique is usually employed. In this procedure a large volume of air is drawn through a filter which traps all particles > 0.01 μm in diameter. The amount retained is weighed and reported (as μg/m^3). Using high-powered microscopes, the collected material can subsequently be identified by size and type, and appropriate size distribution patterns evaluated. [The techniques are described in British Standard B.S.

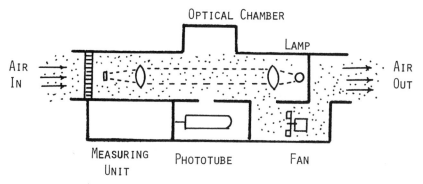

Fig. 3.1 Diagram showing the basic components of a haze meter (after Carmichael, A. J., and Chambers, A. J., Engineering Bulletin ME18, University of Newcastle, 1973).

1747 (1969) and an Environmental Protection Agency Report (Federal Register, 36, No. 84, part II, 1971)].

For studying the finer particles, electron microscopes need to be used. The principles of operation are similar to those of a light microscope, except that the lenses are magnetic fields and an electron beam, rather than a light beam, defines the object. This results in high resolving power, and magnifications of several thousands can be achieved. The enlarged image is either projected onto a fluorescent screen for viewing or on to photographic film for a permanent record.

For the purpose of roughly comparing the total fallout in one area with that of another, some fairly crude procedures often suffice. In one approach a funnel (of known diameter) is placed in the neck of a large bottle mounted about 5 ft above ground in some open space.

Dust falling into the funnel gravitates into the bottle or is washed there by rain. At the end of an extended time period (e.g., one month) the collected material is weighed and subsequently treated to ascertain the proportions which are water soluble or organic in nature. The amounts involved are mg/cm^2 of funnel area but the values are often scaled up to units of $tons/mile^2/month$.

With such small masses of collected material, the measurements are inherently inaccurate, month by month variations can be large, and there is little direct correlation with adverse health effects. At the same time, this simple procedure does permit observation of long-term trends.

B. Specific Particulates

Industrial processes can release very fine particles of a wide range of compounds (e.g., acid mist, asbestos, silica, lime, metal compounds) but few have received the same publicity as the extremely small ($< 1 \mu$m) lead bearing particles produced on combustion of leaded gasoline. In absolute units, the concentrations of lead involved are very small (< 1 ng/m^3 in unpolluted zones, up to 3 μg/m^3 in some city streets, 50 to 70 μg/m^3 on crowded freeways), but the size could ensure almost total retention in the lung.

For average specific particulate values in a given area, large volumes of air are drawn through filter pads over a prolonged time period. A fraction of the pad (e.g., 10%) is dissolved in acid and the filtered extract is examined for several components by standard trace analysis techniques (e.g., atomic absorption spectroscopy). In one city survey, for example, the lead, zinc, and iron contents of suspended matter were all observed to fall in the range of 0.1 to 6 μg/m^3.

For some purposes it is preferable to have instant values of particulate concentration. One proposed method for ascertaining such values uses a sample of the air to support a hydrogen flame. Components of the particulate matter which are sufficiently excited by the high temperature to emit characteristic radiations are evaluated by measuring the intensity of the emissions (i.e., a modification of the technique known as flame photometry, e.g., Section I.B. of Chapter 6).

An alternative approach utilises a microsampling technique. For say, lead determinations, a fixed volume of air (50 to 200 ml) is drawn through a small disc of Millipore filter medium, held in a perforated carbon cup. The cup is subsequently placed between electrical leads and by gradually increasing the applied voltage, the microsample and filter are dried and ashed. A final heavy impulse volatilizes the residue, releasing a puff of metal vapor. If a beam of radiation characteristic of some metal (e.g., Pb), is focused over the top of the cup, the vapor release causes a sharp diminution in intensity. The signal change can be correlated with total metal content through calibration with standards. Picograms (i.e., 10^{-12} g) of metal particles on the disc can yield measurable signals but errors due to contamination or bad technique can be large.

Figure 3.2 shows the basic components of the equipment used. This is essentially an atomic absorption spectroscopy unit, except that a carbon cup and electrodes are used to form atomic vapor in place of the conventional hot gas flame into which solutions are sprayed (shown as

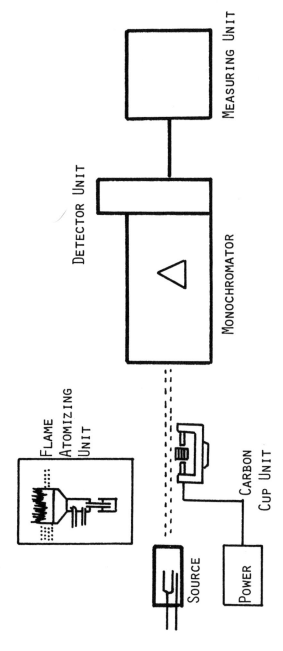

Fig. 3.2 Schematic representation of an atomic absorption spectroscopy unit with carbon cup atomization in place of the standard flame nebulizing unit (inset).

inset). The nebulization of solutions (e.g. from dissolved filters) into a hot flame allows detection and measurement of solution concentrations measured in mg/ml.

II. MONITORING OF CARBON MONOXIDE AND SULFUR DIOXIDE

Motor vehicle exhaust gases contain at least five components which pollute the atmosphere: lead compounds, carbon monoxide, sulfur dioxide, oxides of nitrogen, and unburnt hydrocarbons. The gas composition varies with engine efficiency and mode of operation, e.g., maximum release of oxides of nitrogen occurs during acceleration (up to 3000 ppm) while the maximum content of unburnt hydrocarbons (up to 4000 ppm) is observed during deceleration.

A. Carbon Monoxide

The major toxic component is carbon monoxide. When cars are cruising, the exhaust gases generally contain <1% CO, but when engines are idling or accelerating concentrations can increase to 5% or more.

Carbon monoxide levels in the air are thus much influenced by traffic density and the degree of stopping and starting. Table 3.1 shows

Table 3.1

Variation with Time of Roadside Carbon Monoxide Levels[a]

	ppm Carbon Monoxide		
Time	Day 1	Day 2	Day 3
10 a.m.	9.3	8.0	–
11 a.m.	9.0	11.8	3.6
12 noon	10.4	10.4	6.3
1 p.m.	8.7	8.2	11.1
2 p.m.	8.7	9.3	9.1
3 p.m.	7.9	10.5	5.2
4 p.m.	9.4	10.2	4.6
5 p.m.	12.4	–	7.4

[a]Data provided by Department of Mechanical Engineering, University of Newcastle.

some results obtained in a country shopping area possessing a narrow main road and a succession of traffic control lights. Note that the levels vary each hour. Accordingly, in order to prescribe air quality standards it is necessary to consider both the concentration and the exposure time.

The U.S. Environmental Protection Agency has nominated desirable levels as <9 ppm average for an 8-hr period or <35 ppm for an hourly exposure. (Significant harm to human health may occur when these values are exceeded by a factor of about four.)

Monitoring of carbon monoxide levels thus requires regular recording of results whenever values approach the recommended limits.

One analytic procedure widely used for monitoring purposes utilises the fact that carbon monoxide molecules selectively absorb some bands of infrared radiation.

The gas sample is drawn (by a pump) through a coarse filter into a refrigerated zone (to condense water vapor) and thence into a measuring cell (Fig. 3.3). A similar cell (the reference) is filled with nitrogen,

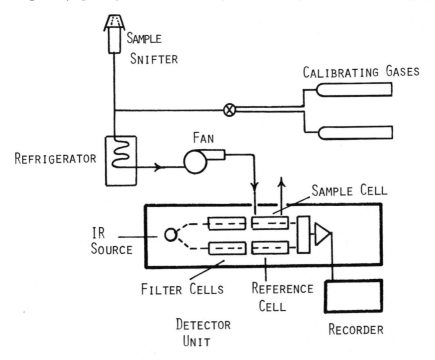

Fig. 3.3 Schematic representation of a carbon monoxide monitor unit (from Engineering Bulletin ME18, University of Newcastle, 1973).

and both are irradiated with infrared radiation derived from a heated
resistance rod. The amount of energy reaching the detector system
from the two cells is unequal if the sample contains carbon monoxide
(due to molecular sorption). This difference is detected, amplified,
and fed to a recorder system.

With mobile, commercial units, errors associated with readings
of about 10 ppm can be of the order of ± 1.5 ppm.

B. Sulfur Dioxide

The sulfur dioxide content of the atmosphere contains contributions
from industry (e.g., acid plants), power stations, and motor vehicles,
and as indicated in Figure 3.4, there can be marked diurnal variations.

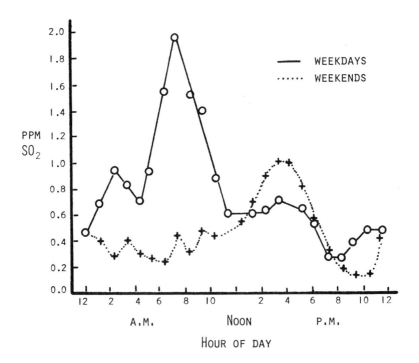

Fig. 3.4 Mean values of sulfur dioxide levels observed at one site over
a period of three weeks. Note the diurnal variation. (Data from
Engineering Bulletin ME18, University of Newcastle, 1973.)

As with carbon monoxide, recommended control levels are therefore based on time averages (e.g., 0.02 ppm/yr; 0.10 ppm/24-hr
period), and to prepare this information it is necessary to make regular
time checks or use a continuous recorder unit.

Figure 3.5 is a schematic representation of one type of unit used
for monitoring sulfur dioxide levels.

Air is drawn into the apparatus (at about 0.01 m^3/hr) through a
filter to remove dust, and then over a heated silver wire to eliminate
substances such as hydrogen sulfide, ozone, and chlorine. The purified gas is then bubbled through an aqueous solution of bromine. Some
bromine is reduced by the sulfur dioxide content and this is continually
replaced by an electrolytic generation process. The current flow during the electrogeneration step can be correlated with the SO_2 content
of the gas. Zero measurements are made by diverting the sample gas
stream through an activated charcoal filter (removes SO_2), and for
calibration a predetermined flow of SO_2 gas is admitted to this purified
stream. If care is taken in calibration and control of flow rates, and
if sorption of SO_2 in the sample line and dust filters can be kept to a
minimum, an overall accuracy of better than $\pm15\%$ may be achieved
using this coulometric procedure (Section VI of Chapter 4).

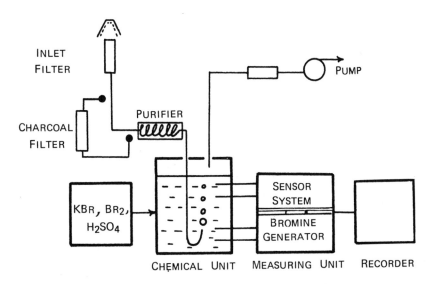

Fig. 3.5 Schematic representation of one type of sulfur dioxide detector (from Engineering Bulletin ME18, University of Newcastle, 1973).

An alternative approach brings the filtered gas sample in contact with hydrogen peroxide solutions. As the sulfur dioxide is converted to sulfuric acid, there is an increase in electrical conductivity which can be readily recorded. This technique is subject to error if the gas sample contains components which dissolve to form electrolytes (e.g., hydrogen halides) or react with acid (e.g., ammonia). For occasional (i.e., check) analyses, one can determine the amount of sulfur dioxide present in a known volume of gas by absorbing it in hydrogen peroxide solution and subsequently determining the acid content by titration or the sulfate ion content by precipitation as barium sulfate.

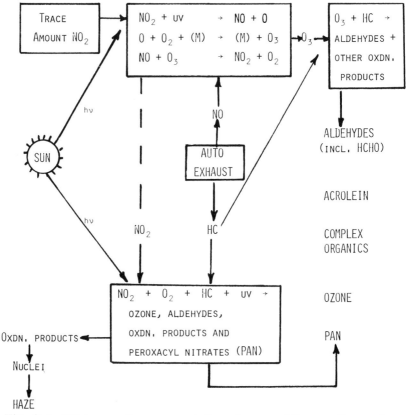

Fig. 3.6 Diagrammatic representation of some of the reaction cycles associated with the formation of photochemical smog.

III. HYDROCARBONS, OXIDES OF NITROGEN, AND SMOG

A. Hydrocarbons in Exhaust Gases

The unburnt hydrocarbons present in engine exhaust gases consist mainly of alkanes, alkenes, and aromatic compounds together with their products of partial oxidation (formaldehyde, acetaldehyde, etc.). Since the concentration of these hydrocarbons reaches a peak with early morning traffic flows, the quality control limit (U.S.) specifies a mean value (0.24 ppm) for a 3-hr period (6 to 9 a.m.). At later hours, sunlight promotes a series of photochemical reactions and leads to the formation of secondary pollutants such as nitrogen dioxide, ozone, aldehydes, ketones, peroxyacyl nitrates (PAN, e.g., $CH_3COO_2NO_2$), and alkyl nitrates. These secondary products give rise to photochemical smog.

Figure 3.6 indicates one proposed sequence of reactions for the formation of photochemical smog. With such a complex mechanism, the concentration of individual components varies widely during the day, hence values such as recorded in Table 3.2 are mainly indicative of relative magnitudes.

Table 3.2

Typical Concentrations of Pollutants in Photochemical Smog[a]

Pollutant	Concentration range (pphm)[b]
Carbon monoxide	200-2000
Total[c] hydrocarbons	20-50[d]
Aromatics	10-30
Aldehydes	5-25
Alkenes	2-6
Nitric oxide	1-15[e]
Nitrogen dioxide	5-20
Peroxyacyl nitrates	1-4
Ozone	2-20

[a]From table in Kerr, J. A., Calvert, J. G., and Demerjian, K. L.: Chem. in Britain 8:252, 1972.

[b]Parts per hundred million.

[c]Excludes methane.

[d]City air hydrocarbon (total) values, 100-800.

[e]City air, average SO_2 values fall in similar range.

The separation and identification of the hydrocarbon species is generally based on gas chromatography or mass spectrometry.

In a gas chromatograph (shown schematically in Figure 3.7), minute amounts of sample mixture are injected into a stream of carrier gas (e.g., nitrogen or helium) flowing through a long tubular column which has been filled with an inert porous powder coated with a nonvolatile oil. Each component in the mixture then flows through the column at a different speed, due to its interaction with the coating on the tube packing. A detector at the other end of the column gives an amplified signal as each component emerges. Under accurately controlled conditions of flow and temperature, each component can be identified on the basis of the time it takes to pass through the column. With careful calibration, the area or intensity of each recorded signal peak becomes a measure of the amount of the component present.

Diversion of the stream leaving the column into a mass spectrometer unit (Fig. 3.8) allows positive identification of the separated components. The individual molecular species are first separated from the carrier gas and are then admitted via a "leak" into a high-vacuum system. In the ionization chamber, the organic materials are exposed to accelerated electrons which on impact convert molecules into charged ions or fragments. All the charged particles are accelerated through electrostatic and magnetic fields and become separated on the basis of their mass/charge ratio. The separated components are sequentially focused on a detector to yield a characteristic mass spectrum. To identify the initial molecular species, this spectrum is compared with that of known pure compounds.

B. Nitrogen Oxides

Nitrogen oxides, NO_x, appear in engine exhausts at concentrations ranging from <30 ppm during idling to >1000 ppm during cruising and acceleration, and reference to Figure 3.6 indicates the important role these oxides play in smog formation, despite their comparatively low concentrations in air [e.g., recommended limit (U.S.) 0.055 ppm/yr].

Nitrogen dioxide contents may be determined by "scrubbing" this component out of an air sample with a reactive solution capable of converting it into a highly colored species. The color-forming reaction (carried out in a continuous flow, counter-current absorption column) usually involves diazotization (e.g., of sulfanilic acid) followed by coupling to an aromatic amine (e.g., N-(1-naphthyl)-ethylenediamine dihydrochloride). The column discharge is fed into a flow-type measuring cell so that the intensity of the color can be determined using some form of spectrophotometer.

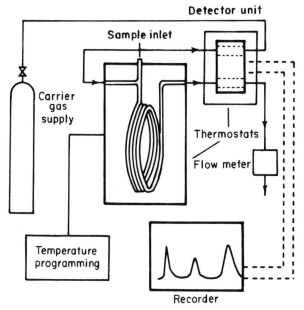

Fig. 3.7   The components of a typical gas chromatograph.   (Reproduced from Pickering, W. F.: "Modern Analytical Chemistry," Dekker, 1971 with permission of the publisher.)

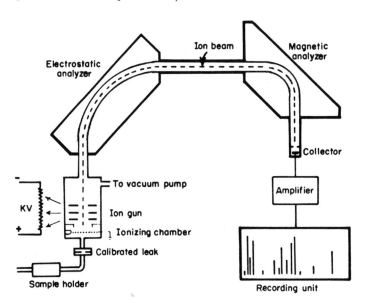

Fig. 3.8   Schematic diagram of a mass spectrometer.   (Reproduced from Pickering, W. F.: "Modern Analytical Chemistry," Dekker, 1971 with permission of publisher.)

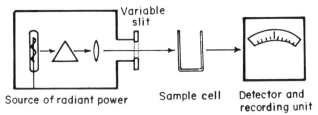

Fig. 3.9 The basic components of an absorptiometer unit. (Reproduced from Pickering, W. F.: "Modern Analytical Chemistry," Dekker, 1971 with permission of publisher.)

A spectrophotometer consists basically of a monochromator, isolating slits, an absorption cell, and a detector system (Figure 3.9). In the monochromator, white light (or ultraviolet light) is dispersed by a prism or grating, and by rotating the dispersive medium, different wavelengths are brought to focus on the exit slit. For a given solution, held in the absorption cell, some wavelengths are absorbed more strongly than others. To make quantitative measurements, a narrow beam of the characteristic wavelengths (i.e., maximum absorption region) is passed alternatively through a solvent blank and the test solution. Absorption by the test solution decreases the amount of light reaching the photoelectric detector. The resultant change in meter reading is then calibrated in terms of the amount of absorbing species present.

Since the efficiency of the color-forming process can vary between columns, the procedure should be calibrated by means of standard gas mixtures. But since handling nitrogen peroxide for dynamic calibration operations can prove difficult, stock solutions of sodium nitrite are more regularly used for routine calibrations. However, with the latter standard one should first establish the appropriate empirical relationship between nitrite concentration and nitrogen dioxide values. One widely assumed factor is 0.72, that is, one mole of the gas produces the same color intensity as 0.72 mole of nitrite.

The nitric oxide content of gas samples may be determined by first oxidizing it to the dioxide, e.g., by passage through acidified potassium permanganate solution. A more sensitive procedure is that based on molecular chemiluminescence.

The gas sample is mixed with freshly generated ozone, to yield "excited" nitrogen dioxide.

$$NO + O_3 \rightarrow NO_2^* + O_2$$

Normal deactivation is accompanied by radiation emission.

$$NO_2^* \rightarrow NO_2 + h\nu \text{ (light)}$$

The intensity of the emitted light is measured with a sensitive detector. The apparatus is calibrated against standard gas mixtures.
This can introduce some errors because the efficiency of the basic
process is determined, in part, by the number and type of quenching
species (M) present,

$$NO_2^* + M \rightarrow NO_2 + M$$

and it is difficult to ensure complete similarity between assays and
standards.

IV. MINOR COMPONENTS

While the pollutant species considered in the preceding sections receive
the greatest amount of general observation, minor components can have
greater impact on physical well being.

These minor components can be subdivided into broad categories,
such as industrial toxicants (e.g. solvent vapors), corrosive agents
(acid vapors), phytotoxicants (SO_2, ethylene, ozone, fluorides), lachry-
mators (aldehydes, acrolein, PAN), allergens, alkylating agents,
pesticides, and carcinogens (arenes, alkanes, phenols).

The physiologic significance, and modes of study, of a wide range
of these minor aerotoxicants has been the subject of a review [Sawicki,
E.: Crit. Revs. Anal. Chem. 1:275, 1970] and this information is
continually updated in specialist publications.

The techniques required to examine these materials are varied.
For example, the ozone content of gases has been determined by absorbing it in potassium iodide reagent (buffered to pH 7) in a continuous-
flow counter-current absorption column. The ozone forms an equivalent
amount of iodine, and the concentration of this product may be evaluated in several ways, including measurement of the color intensity with
a spectrophotometer. Positive errors are introduced by compounds
such as nitrogen dioxide (10 ppm NO_2 having similar effect to 1 ppm
O_3) and negative errors are introduced by the presence of reducing
agents (e.g., SO_2).

Many of the aerotoxicants are determined spectrophotometrically,
that is, they are caused to undergo specific reactions with selected

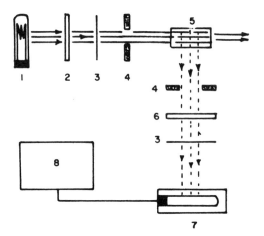

Fig. 3.10 Schematic representation of a fluorimeter: 1. Radiation source, 2. excitation filter or monochromator, 3. shutters, 4. variable apertures, 5. sample cell, 6. emission filter or monochromator, 7. phototube detection unit, and 8. recorder or meter. (Reproduced from Pickering, W. F.: "Modern Analytical Chemistry," Dekker, 1971 with permission of publisher.)

reagents to yield products which absorb strongly in some part of the visible or ultraviolet regions of the spectrum.

An alternative approach (which is often more sensitive) involves conversion of the species of interest into a fluorescent compound. When excited with ultraviolet light, the latter emit characteristic visible radiations. The intensity of these emissions is measured by mounting a photomultiplier detector at right angles to the exciting beam (as shown in Fig. 3.10).

$$SR + h\nu \rightarrow SR^* \rightarrow SR + h\nu' + heat$$
(UV light)   (visible light)

The types of substances which fluoresce are mainly aromatic organic compounds (e.g., benzene, naphthalene, anthracene, and their derivatives) or a metal-fluorogenic reagent complex.

With careful selection of excitation wavelength and optimization of conditions, concentrations as low as 0.01 ppm may be determined. On the other hand, great care is required to prevent contamination of the sample, for example, common solvents often contain fluorescent substances, stoppers can contain extractable fluorescent materials, and

grease or paper fibers can introduce interfering species. In every set
of determinations, the analyst is well advised to run a blank and at least
two standards of known composition covering the concentration range
of interest. With ideal systems, the intensity of emission is directly
related to the concentration of fluorescing material.

As in the molecular chemiluminescence technique for nitric oxide,
the accuracy depends greatly on the absence of species capable of
quenching, that is, promote deactivation of the excited species without
the liberation of light.

For mixtures of aerotoxicants, separation of the components by
some form of chromatography is usually adopted. Gas chromatography
is very useful as a separation procedure and for quantitative measurements, but as a tool for the identification of species it has limitations.
This deficiency can be overcome through the use of accessories which
permit collection of the various separated fractions. The latter can
then be submitted to examination in a mass spectrometer or an infrared spectrophotometer.

An infrared spectrophotometer allows one to obtain molecular
fingerprints which can be used to identify compounds with a high degree
of certainty. The instrument contains a source of infrared radiation
(a heated bar), a dispersing medium (e.g., NaCl prism), a sample cell,
and ir detector (Fig. 3.11). By moving the dispersing medium, the

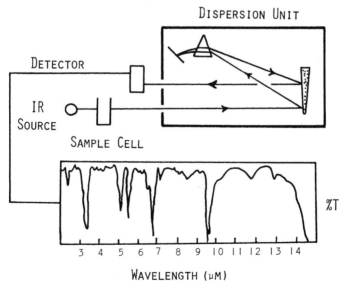

Fig. 3.11 Diagrammatic representation of an infrared spectrometer
and fingerprint spectrum.

sample can be exposed to a wide spectrum of infrared radiations. Different parts of molecules exhibit differing interatomic vibrational movements, and the amplitude of these vibrations can be increased through absorption of specific bands of infrared radiation. As a result, when the dispersed infrared radiations are successively passed through the sample, varying proportions are transmitted and the graph produced shows the extent of absorbance at specific wavelengths. The wavelengths at which absorption peaks occur are characteristic of particular types of atomic groupings, but the combined effect for a pure organic compound is exclusive to that compound, i.e., it is an ir fingerprint. If only one absorbing species is present, positive identification can be achieved through comparison with the spectra of pure, known compounds.

Since the concentrations present in gas samples are normally extremely small, it is necessary to fit the infrared spectrophotometer with an accessory which increases the effective path length. For radiation absorption, the relationship between concentration (c) and effective path length (b) is of the form

$$\text{Absorbance} = a \cdot b \cdot c = \log \frac{P_0}{P}$$

where P_0 is the intensity of the incident radiation beam, P is the intensity after absorption by sample components, and a is a factor characteristic of the species. Using this relationship, a wavelength of radiation which is strongly sorbed, and a values determined by calibration with standards, one can make a quantitative evaluation of the amount of pure substance present.

From a study of the absorption spectrum (usually covering the wavelength range 2 to 16 μm) of a normal air sample, one can identify peaks which indicate the presence of particular chemical groups in the sample (e.g., $-C=$); $-NH_2$; $-CH=CH-$). Almost equally important is the negative information that can be derived. For example, if no strong absorption bands are observed between the wave numbers 1600 and 1800 cm^{-1} (wavelength 6.26 and 5.55 μm), the presence of carbonyl groups can be ruled out and this eliminates from consideration all ketones, aldehydes, organic acids, esters, and similar compounds. The absence of bands above 3200 cm^{-1} (i.e., wavelength $< 3\ \mu$m) permits elimination of the presence of alcohols, amines, amides, and other substances.

The need for efficient separation procedures may become more apparent if one considers the composition of a material which has been widely investigated in recent years, namely, cigarette smoke. The solid components of the smoke are mainly tars and nicotine (an alkaloid) with some suspected carcinogenic compounds (such as benzopyrin and

catechol) present as minor components. In the gas phase a wide range
of compounds have been identified, including carbon monoxide (16,000),
acetaldehyde (1000), acetic acid (600), formic acid (500), nitric oxide
(400), hydrogen cyanide (300), nitrogen dioxide (200), acetonitrile (140),
phenol (120), ammonia (100), butadione (50), formaldehyde (40), and
hydrogen sulfide (10). The values in parentheses indicate a relative
or average concentration (expressed in the units μg/cigarette).

Besides the techniques mentioned in the preceding examples, others
such as microtitrations or electrometric measurements are in widespread use. For example, fluoride vapors are scrubbed from gases
with an alkaline solution, separated from interferent species by selective distillation, and finally determined by means of an ion-selective
electrode (discussed in Chapter 6).

For accurate, quantitative measurements of the trace components
present in a given gas sample, it is necessary to have skilled operators,
high-quality equipment, and efficient calibration. Unfortunately, the
development of sensitive detectors for trace atmospheric pollutants has
far surpassed the facility to calibrate them accurately.

In the past, standard gas mixtures (in the ppm range) were most
conveniently prepared by mixing an aliquot of pure component with a
large volume of diluent gas. The reliability of this method depends on
the accuracy with which aliquot and diluent may be measured, the
mixing compatibility of the components, and the magnitude of any surface adsorption losses, or chemical interaction effects.

A newer method makes use of the ability of gases or vapors to
permeate through Teflon tubing. The rate of permeation is a function
of temperature, length and diameter of tube, and wall thickness; hence
by operating under standard conditions, constant amounts of sample
vapor can be added to a stream of diluent gas.

The analytical problems of air pollution monitoring are continually
being resolved, and the errors in most cases are far below the variance
of sample collection.

It is perhaps appropriate to conclude this brief introduction to air
monitoring, with mention of a technique which has proved very useful
for routine monitoring of a range of industrial vapors or toxicants, i.e.,
specific absorption tubes.

By means of a hand pump, known volumes of the atmosphere are
drawn through a tube filled with a special permeable absorbent mixture.
The absorbent chosen changes color as chemisorption of a particular
component occurs, and the amount of pollutant is estimated from the

length of the discolored segment. For example, one adsorbent used
in alcohol-in-breath studies is acidified potassium dichromate mounted
on silica gel. Oxidation of adsorbed alcohol by this reagent leaves a
distinct green band. Instead of a tube, pads of paper soaked in appropriate chemicals can be used, with measurements being based on color
intensities.

This mode of monitoring gives instantaneous, approximate values
which are adequate for ensuring that dangerous thresholds or permitted
limits are not being grossly exceeded.

FURTHER READING

American Public Health Association, "Methods of Air Sampling and
Analysis," Washington, 1972.

"Annual Reviews," Anal. Chem., No. 5 (April), Vol. 43 (1971); Vol.
45 (1973); Vol. 47 (1975).

Brenchley, D. L., Turley, C. D., and Yarmac, R. F.: "Industrial
Source Sampling," Ann Arbor Science Publishers, Ann Arbor, 1973.

"Chromatography of Environmental Hazards,"
      Vol. I Carcinogens, Mutagens and Teratogens.
      Vol. II Metals, Gaseous and Industrial Pollutants.
Elsevier, Amsterdam, 1972, 1973.

Hesketh, H. E.: "Understanding and Controlling Air Pollution," Ann
Arbor Science Publishers, Ann Arbor, 1972.

Katz, M.: "Measurement of Air Pollutants," World Health Organiza-
tion, Geneva, 1969.

Kolthoff, I. M., Elving, P. J., and Stross, F. H., eds.: Part III,
Vol. 2, Industrial Toxicology and Environmental Pollution and its
Control. Wiley, N.Y., 1970.

Ledbetter, J. O.: "Air Pollution Pt. A. Analysis." Dekker, N.Y.,
1972.

Leithe, W.: "The Analysis of Air Pollutants," Ann Arbor-Humphrey
Science Publishers, Ann Arbor, 1970.

Magill, P. L., Holden, F. R., and Ackley, C., eds.: "Air Pollution
Handbook," Chapters 10 and 11. McGraw Hill, N.Y., 1956.

McCormac, B. M., ed.: "Introduction to the Scientific Study of
Atmospheric Pollution," Reidel Publishing Co., Dordrecht-Holland, 1971.

Ruch, W. E.: "Quantitative Analysis of Gaseous Pollutants," Ann Arbor-Humphrey Science Publishers, Ann Arbor, 1970.

Stern, A. C., ed.: "Air Pollution," 2nd ed.
    Vol.   I Air Pollution and Its Effects;
    Vol. II Analysis, Monitoring and Surveying;
    Vol. III Sources of Air Pollution and Their Control;
Academic Press, N.Y., 1968.

Stevens, R. K. and Herget, W. F., eds.: "Analytical Methods Applied to Air Pollution Measurements," Ann Arbor Science Publishers, Ann Arbor, 1974.

4

PRINCIPLES OF GRAVIMETRIC, TITRIMETRIC, AND ABSORPTION METHODS

Weight measurements have been used as a basis for quantitative chemical analysis for over a century, and while the popularity of such procedures has declined in the last 20 years they still possess value as check procedures. Unlike most instrumental methods which rely on comparing one physicochemical response with another, gravimetric procedures are direct and absolute. The time required for analysis and the concentration ranges involved limit their applicability in pollution evaluation, but the principles are quite relevant, since precipitation and selective sorption are used for preconcentration and separation of system contaminants.

Titrimetry has provided satisfactory means of determining species in solution for a period nearly as long as have weight measurements. Over the years, there have been improvements in end-point detection and changes in the nature of the species used as the titrant. These refinements have allowed the technique to be extended into regions of lower concentrations (bordering on those encountered in pollution studies) and have facilitated semiautomation of the process. For routine studies of a selected range of chemical species, titrimetry remains the preferred method.

Absorption methods have developed rapidly since the widespread introduction of commercial absorptiometers in the post-1950 period. As noted in the preceding chapter, this technique has many applications in pollution evaluation studies and can be regarded as one of the basic tools required for this type of investigation.

I. GRAVIMETRIC METHODS

The fundamental principles of gravimetric analysis, based on precipitation, are extremely simple: select a chemical reaction that yields a sparingly soluble product, isolate, and weigh the product; e.g.,

$$SO_4^{2-}{}_{(aq)} + Ba^{2+}{}_{(aq)} \rightleftarrows BaSO_4{}_{(s)}$$

The time required for gravimetric analysis can be considerable, but as modern analytical balances weigh to a fraction of a milligram, with due care in the separation stage, accuracies equivalent to a fraction of a percent can be regularly obtained. The technique is absolute, hence it is often used to analyze the standard samples required for calibration of the comparative techniques described in other sections. To achieve the accuracy inherent in the method however, an adequate understanding of the factors that influence the solubility and purity of precipitates is essential.

The chemical behavior of many elements is similar, and accordingly a major limitation of the gravimetric technique is a general lack of specificity. In addition, the majority of solid reaction products are ionic compounds which are subject to marked adsorption effects. These limitations have been minimized by introducing organic precipitating reagents.

A. Solubility Product

The distribution of a sparingly soluble compound, A_xB_y, between the solid and aqueous phases can be represented as:

$$A_xB_y{}_{(solid)} \rightleftarrows A_xB_y{}_{(aq)} \rightleftarrows xA^{y+} + yB^{x-}$$

The equilibrium constant for such reactions is known as the solubility product (S) and

$$S_{A_xB_y} = (a_A)^x(a_B)^y = [A^{y+}]^x[B^{x-}]^y \cdot f_A^x \cdot f_B^y$$

a_A is the activity of species A, [A] is its molar concentration, and f_A its activity coefficient.

Much of the observed variable solubility of ionic solids in electrolyte solutions can be attributed to variations in the values of the activity coefficients (f_A and f_B). These values vary with the charge, size, and concentration of all ionic species in solution and approach a value of unity only in extremely dilute solutions.

In <u>gravimetric analysis</u>, the aim is to remove one ion (e.g., A^{y+}) almost completely from the solution by causing it to be present mainly in the solid phase. The term $[A^{y+}]$ is thus a measure of the solubility of the precipitate, and for quantitative studies, $[A^{y+}]$ must be as small as possible. To achieve this aim, $[B^{x-}]$, f_A, and f_B must be as large as possible.

An addition of excess common ion B^{x-} increases $[B^{x-}]$ and reduces the solubility, but there is a limit to the amount of excess that can be added, because a point is reached where the influence of the excess on the activity coefficients is greater than the benefit gained from the large value of $[B^{x-}]$.

The removal of A^{y+} or B^{x-} ions from the solution by competing equilibria reduces the amount of A^{y+} in the solid phase and the solubility of the precipitate is accordingly increased. For example, if the anion B^{x-} is derived from a weak acid, then in solution the equilibrium relationship

$$K_a = \frac{[H^+]^x [B^{x-}]}{[H_x B]}$$

has to be satisfied, and it can be seen that $[B^{x-}]$, hence $[A^{y+}]$, depends on $[H^+]$ (i.e., the pH of the solution).

The assay solution may also contain species that form complexes with A^{y+}, e.g.,

$$A^{y+} + nL \rightleftarrows AL_n^{y+}; \quad K = \frac{[AL_n^{y+}]}{[A^{y+}][L]^n}$$

In this case, the amount of A in solution $([A^{y+}] + [AL_n^{y+}])$ depends on the relative magnitudes of $S_{A_x B_y}$ and K, and the relative amounts of B^{x-} and L (both functions of pH).

In summary, the solubility of a precipitate is influenced by the pH of the solution, by the presence of complexing agents, and by the ionic strength of the solution.

B. Purity of Precipitates

Even with minimum losses due to solubility, high accuracy can be achieved only if the isolated material is virtually 100% pure, can be wholly retained by the filter medium, and can be washed free of all adhering nonvolatile impurities from the original solution.

The fulfilment of these ideals depends largely on the particle size of the precipitate.

For any precipitate, a minimum number of ions or molecules must become associated in order to produce a stable second phase. This minimum-sized particle is called a nucleus, and it has been proposed that the number of nuclei formed under given conditions (N) follows a relationship such as

$$N = \frac{K\ (Q - S)}{S}$$

where K is a constant, Q is the concentration of precipitate components in solution at a given instant, and S is the equilibrium concentration in a saturated solution, (Q - S) is known as the degree of supersaturation.

The second process that occurs during precipitation is the growth of particles already in the solution. The rate of growth has been found to be proportional to the degree of supersaturation and the surface area of the exposed solid, but if conditions favor the formation of a large number of nuclei, little material is available for growth, and particle size of the final product can be extremely small (e.g., 0.001 to 0.1 μm).

It can be observed from the above equation that the number of nuclei formed may be minimized by increasing S (e.g., by a change in pH, increase in temperature) or decreasing Q (e.g., by adding dilute reagent with adequate stirring).

In some cases the effect of varying Q and S is insignificant and the final product consists of a mass of particles of colloidal size. In other cases, altering these two factors markedly reduces the number of nuclei formed and the final precipitate is composed mainly of large particles. Large particles are sought, because they are easier to filter and wash, and are purer (due to slower growth and smaller amount of adsorbed material).

As shown schematically in Figure 4.1, particles grow by the addition of positive and negative ions on all available surfaces. The process continues until the concentration of one of the ions approaches zero. At this stage, the second ion proceeds to take up its normal lattice position to yield a surface that is electrically charged due to absence of the second component. The monolayer on the surface is known as the primary adsorbed layer and the electrical charge tends to be neutralized by a diffuse zone of oppositely charged ions from the bulk solution.

The extent to which precipitates are contaminated by surface-adsorbed material depends on the surface area per gram and the affinity of diverse ions for the species present in the primary adsorbed layer. The amount sorbed per gram (x/m) can be related to the concentration of diverse electrolyte (c) by an empirical mathematical relationship

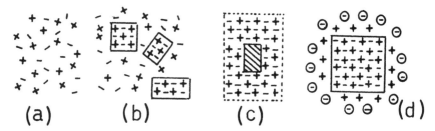

Fig. 4.1 Diagrammatic representation of processes involved in crystal growth, + represents a cation and - represents an anion capable of combining to form a sparingly soluble compound. (a) Supersaturated solution; (b) nuclei formation; (c) particle growth; (d) adsorption on crystal surface, showing primary layer of cations +, and diffuse secondary layer of "foreign" anions $\ominus$. (Reproduced from Pickering, W. F., "Fundamental Principles of Chemical Analysis," Elsevier, 1966 with permission of publisher.)

such as

$$\frac{x}{m} = kc^n$$

where n is <1 and k varies with the size of the particles and the nature of contaminant.

With particles of colloidal size, the amount sorbed can be quite significant. However, if all electrolyte is removed, the material disperses, because the charge due to the primary adsorbed layer causes the tiny particles to repel each other. To purify coagulated colloidal precipitates (such as the hydrous oxides of iron, aluminium, etc.) it is thus necessary to collect the initial product, then redissolve and reprecipitate the material in the presence of secondary ions which can be lost on subsequent heating (e.g., NH_4NO_3).

During the growth of a crystal, all charged ions in solution tend to come in contact with the developing surfaces. With slow conditions of growth, foreign ions have time to diffuse away from the surface and growth takes place in the ideal manner indicated in Figure 4.1. However if the foreign ion forms a strong bond with an ion of the precipitate, and the ionic sizes are comparable ($\pm 10\%$), then the foreign ions may become integral parts of the crystal lattice. This process is known as solid solution and it is extremely difficult to free a precipitate from such impurities. Examples of solid solutions are $PbSO_4$-$PbCrO_4$, $PbSO_4$-$BaSO_4$, $MgNH_4PO_4$-$MgKPO_4$ mixtures.

With conditions favoring rapid growth, foreign ions of many dimensions can be trapped inside the growing crystal. The trapped ions are said to be <u>occluded</u> impurities and their presence not only reduces the purity of the precipitate but it induces faults in the crystal surface, and growth becomes irregular and uneven in different directions.

With crystalline precipitates of appreciable particle size (order of microns), purification may be achieved by digestion of the precipitate in contact with its mother liquor at an elevated temperature for a period of time (e.g., 1 hr). Small particles and sharp extremities, tend to dissolve in preference to larger, more regular crystal surfaces. The excess ions from solution, however, tend to deposit on the more regular surfaces. This slow solution and deposition process frees occluded impurities and the digested material is usually purer, larger, and more regular in shape than the initial product.

In some cases, the impurity deposits on the surface of the original precipitate subsequent to quantitative deposition of the latter, and the amount formed increases with time. Common examples are the post-precipitation of zinc on copper sulfide and magnesium on calcium oxalate. The phenomena known as post-precipitation can be explained in terms of the concentration of anions at the surface of the original precipitate being of such a magnitude that the solubility product of the second compound is exceeded in the region of this surface layer.

After purification (by digestion or re-precipitation) precipitates have to be isolated from the mother liquor. The filter medium used for this purpose should be of the minimum porosity required to retain the material and may be either a sintered glass or porcelain crucible, or a filter paper. The transfer of precipitate to the filter medium must be quantitative, and has to be followed by an efficient washing process.

The wash liquor must not dissolve the precipitate but it should displace the impurities attached to the surface and replace them with species that are completely removed in subsequent heating processes. The nature varies from precipitate to precipitate. For example, barium sulfate may be washed with water, dilute ammonia solution is used for magnesium ammonium phosphate, for hydrous iron oxide ammonium nitrate solution is recommended, and dilute nitric acid is suitable for washing silver chloride.

After washing, the precipitates have to be dried and converted to a form suitable for weighing. The product weighed must have a definite, known, chemical composition and should preferably be nonhygroscopic. In some cases heating to 105°C to remove water is adequate; at other times temperatures much greater than $500^\circ C$ are needed to destroy filter paper, or other organic matter, and achieve required chemical transformations. The amount of heating required is best determined by thermogravimetric studies.

In thermogravimetry, a sample is simultaneously heated and weighed; the weight being plotted as a function of temperature. In regions where the product has been converted into a stable chemical form, no weight loss is observed over an extended temperature range, and this range can prove highly suitable for routine drying or ignition of analytic precipitates of the same material.

C. Selectivity

Inorganic reagents are generally not very selective when used to form sparingly soluble compounds, and hence isolated species can be badly contaminated with co-precipitated material if several ionic species are present in the same solution. For example, the hydrogen phosphate ion (HPO_4^{2-}) can be employed for the quantitative determination of Bi, Co, Zn, Zr, Be, Mg, and Mn; also precipitated are the ions Hf, In, Ti, Au, Ba, Ca, Hg, La, Pb, rare earths, Sr, Th, and U. Similarly, in dilute nitric acid media the addition of silver ions leads to precipitation of at least 12 anions (Br^-, CN^-, SCN^-, I^-, etc.).

The selectivity of gravimetric procedures can be increased by utilizing organic reagents. Organic reagents tend to react selectively with a limited number of elements, and the reaction can often be made specific by the control of pH and by the addition of other complexing ligands to the solution.

Because the precipitates are not ionized salts, they do not co-precipitate impurities in the same way as barium sulfate and other ionic precipitates. On the other hand, organic reagents are usually insoluble in water and have to be added as a solution in another solvent. Too large an excess can thus lead to contamination of the metal chelate with precipitated reagent. This problem can be overcome by igniting the precipitate at a temperature which destroys all organic matter (e.g., leaving residue of metal oxide), but such treatment nullifies another important advantage of organic reagents.

Owing to the high molecular weight of the reagents, a little metal ion yields a large amount of precipitate. Accordingly, the technique is very suitable for isolating low concentrations of cations, provided the material can be dried (e.g., at $105^\circ C$) and weighed as precipitated.

The selectivity of organic reagents is associated with the presence (and location) of particular atomic groupings in the reagent molecule.

The organic reagents that react with metal ions all contain groups with replaceable hydrogen atoms (-COOH, -SO$_3$H, -SH, etc.). Reagents which form insoluble chelate compounds contain, in addition, a functional group of basic character such as -NH$_2$, =N-, =O, with which the reacting metal is coordinated to form a five- or six-membered ring.

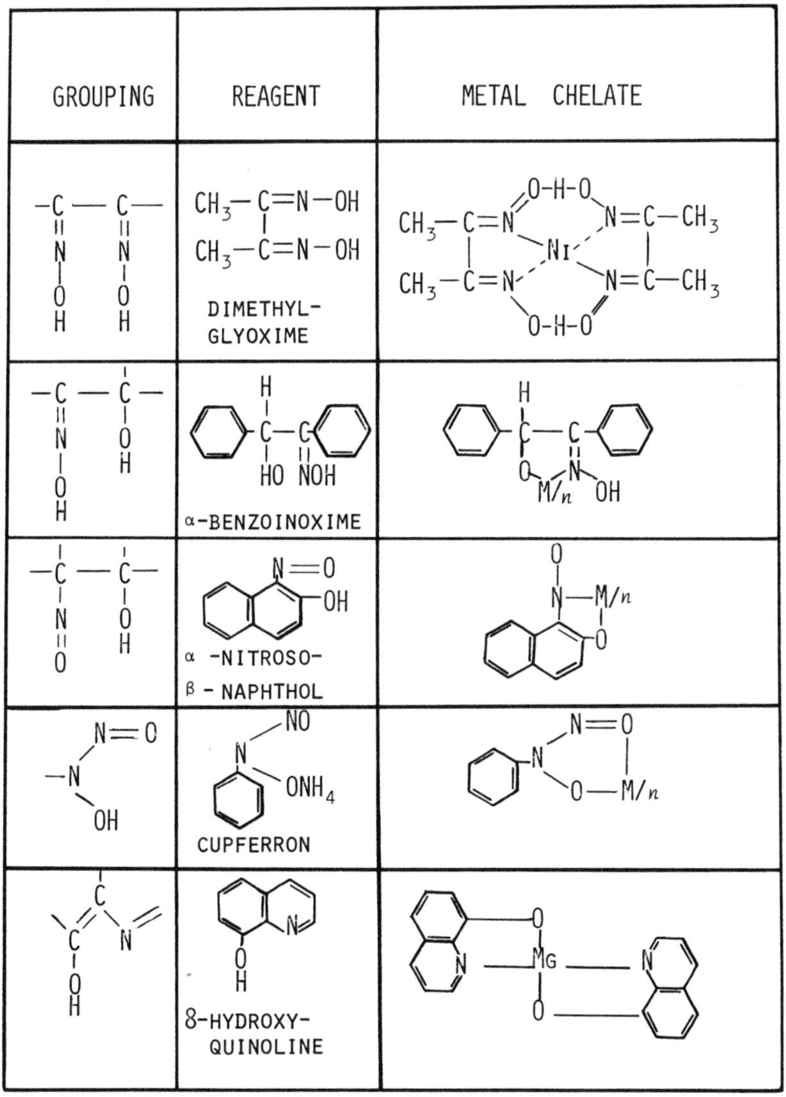

GROUPING	REAGENT	METAL CHELATE
-C—C- (both C double-bonded to N, N–O–H)	CH$_3$—C=N—OH CH$_3$—C=N—OH DIMETHYL-GLYOXIME	Nickel dimethylglyoxime chelate
-C—C- (C=N–O–H and C–O–H)	C$_6$H$_5$—CH—C=NOH (HO NOH) α-BENZOINOXIME	Metal α-benzoinoxime chelate
-C—C- (N=O and O–H)	α-NITROSO-β-NAPHTHOL	Metal α-nitroso-β-naphthol chelate
—N(N=O)—OH	CUPFERRON (N–NO, ONH$_4$)	Cupferron metal chelate
-C=C-N= with C–O–H	8-HYDROXY-QUINOLINE	Magnesium 8-hydroxyquinoline chelate

Fig. 4.2 Typical functional groups found in organic chelating agents.

Five typical reactive groupings are shown in Figure 4.2 together with common reagents containing these groupings. The manner of metal bonding in chelates is indicated by the structures shown for magnesium 8-hydroxyquinolate and nickel dimethylglyoximate on the same diagram.

The acidic and basic groups must be so spaced that they can form a chelate ring, which is almost completely free of strain. Not only must the organic molecule be of the right size and shape but the metal ion must also have the right size, the right oxidation state, and the right coordination number.

The pH is always important in precipitations with complex-forming organic reagents, because these are all weak acids, and the higher the pH, the more readily are their hydrogen atoms displaced, e.g.,

$$Mg^{2+} + 2HOx \rightleftarrows Mg(Ox)_2 + 2H^+$$

where HOx = 8-hydroxyquinoline.

Substituent groups in the organic molecule can influence the acidity and basicity of the two functional groups, and other substituents can influence the solubility of the metal chelate. Adding hydrocarbon substituents generally decreases the solubility of the precipitate and reagent in water, while the opposite effect can be observed by introducing polar groups such as $-OH$ and $-SO_3H$.

Gravimetric procedures are at a disadvantage when the amounts sought weigh less than a milligram, hence selective precipitation is rarely used in trace analysis studies. However, in this range the selectivity of organic reagents still can be utilized, since the metal chelates are soluble in organic solvents and the extracts are generally sufficiently highly colored to allow their use in spectrophotometric evaluations.

II. ACID-BASE TITRATIONS

Volumetric titrations provide a simple and convenient means of analyzing for species in solution.

In this technique, a solution of one reactant (of known strength) is added from a burette to a solution containing the species to be determined. Provided that the reaction between the added solution and the species to be determined proceeds rapidly to completion, and it is possible to express the chemical reaction in the form of a stoichiometric equation, the volume of reactant added can be used to determine the composition of the unknown.

In order to utilize the titrimetric procedure it must be possible to detect the end-point in the reaction. This problem is simplified by the fact that there is a rapid change in the concentration of some component of a reaction in the vicinity of the equivalence point.

A number of chemical compounds respond to this rapid change in concentration by changing some visible property (e.g. color, solubility) and such compounds are called indicators.

The selection of a suitable indicator for a given titration requires a knowledge of the equilibrium conditions in the solution at all stages of a titration, a knowledge of the mechanism of indicator behavior, and an understanding of solvent effects.

A. Titration Curves

If an acid, HA, is titrated with a base, B, a plot of pH (i.e., $-\log [H^+]$) versus volume of titrant added generally yields an S-shaped curve, commonly known as a titration curve (Fig. 4.3).

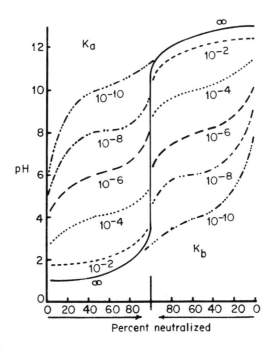

Fig. 4.3 Variation in pH observed on titration of 0.1 M acids (or bases) possessing different dissociation constants. The curves for base titration (on right-hand side) have been plotted in the reverse direction so that by combination of two appropriate segments one can derive a titration curve. (Reproduced from Pickering, W. F.: "Modern Analytical Chemistry," Dekker, 1971 with permission of publisher.)

The curve can be considered to be composed of two parts, joined at the point of inflexion (which corresponds to conditions at the equivalence point in the titration). The slope of the initial (excess acid) and final (excess base) segments, and the position of the inflexion point (on the pH scale), are controlled by the concentrations of all the species present at particular stages of the titration and the magnitude of the equilibrium constants associated with solute-solvent interactions.

In solutions containing excess acid, the concentration of hydrogen ions (hence pH) is related to the position of an equilibrium situation such as:

$$HA + H_2O \rightleftarrows H_3O^+ + A^- ; \quad K_a = \frac{[H^+][A^-]}{[HA]}$$

K_a values vary with the nature of the acid involved and the temperature; the ratio $[HA]/[A^-]$ varies during the course of the titration. If enough base has been added to neutralize 1% of the total acid, $[HA]/[A^-] = 99/1$; neutralization of half the acid leaves equal amounts of HA and A^-, and $[H^+] = K_a$; when only 1% of acid is left, $[HA]/[A^-] = 1/99$.

Figure 4.3 shows the effect of different K_a values on the pH changes during neutralization.

A number of acids ionize completely (represented by $K = \infty$) and the concentration of hydrogen ions in solution can then be taken as numerically equal to the concentration of acid present. Titration of these strong acids (HCl, HNO_3) with a base reduces the amount of acid remaining, and dilutes the solution, hence when $K_a = \infty$,

$$H^+ = \frac{(\text{Volume of unreacted acid}) (\text{initial molarity})}{(\text{Total volume of solution})}$$

[If 100 ml 0.1 M HCl are treated with 10 ml 0.2 M NaOH, 80 ml of the original acid will remain unreacted, and $[H^+] = (80) (0.1)/110$ or pH = 1.14 (initial pH of 1.0)].

Note that with strong acids, the initial pH (hence whole segment location) varies with the initial acid concentration (e.g., 0.001 M HCl, pH = 3).

In solutions containing free base (e.g., after equivalence point in the above titration), the pH is controlled by the amount and type of base present:

$$B + H_2O \rightleftarrows HB^+ + OH^- ; \quad K_b = \frac{[HB^+][OH^-]}{[B]}$$

In a titration with base, the ratio $[B]/[HB^+]$ increases as more and more excess base is added, and the $[OH^-]$ (hence pH) of the resulting solution becomes a function of the stage of titration and the magnitude of K_b (as shown in Fig. 4.3).

Where bases ionize completely in water, (e.g., NaOH, KOH), $K_b = \infty$, and $[OH^-]$ = (volume of base unreacted) $\cdot$ (initial molarity)/ total volume.

For titrations in which a single proton is transferred, the total titration curve resembles the shape obtained by bringing together appropriate K_a, K_b segments from Figure 4.3. Thus if 0.1 M HCl ($K_a = \infty$) is titrated with 0.1 M NH_3 ($K_b = 10^{-5}$), then the pH jump near the equivalence point is from about 3 to 8. Similarly in the titration of a weak acid (e.g., $K_a = 10^{-6}$) with 0.1 M NaOH, one would expect the pH to change from about 8 to 11 as just enough base is added to completely neutralize the acid initially present.

One may generalize by stating that as the values of K_a or K_b for the reactants become numerically smaller, the change in pH near equivalence point is smaller and indicator selection is more critical (as discussed in the next section).

Calculated pH values tend to differ from experimental values, because concentrations rather than activities are substituted in the equilibrium equations, but considerations such as the above are useful whenever it is not possible to establish the form of titration curve directly, (i.e., by pH meter).

The experimental determination of titration curves is the preferred method of investigating the behavior of new titration systems. It is widely applied to detect the equivalence point in titrations involving colored or opaque solutions, and in reactions for which there is no suitable visual indicator.

In titrations of polybasic acids (e.g., H_3A) the protons are removed in a stepwise fashion

$$H_3A \overset{K_1}{\rightleftarrows} H_2A^- + H^+; \quad H_2A^- \overset{K_2}{\rightleftarrows} HA^{2-} + H^+; \quad HA^{2-} \overset{K_3}{\rightleftarrows} A^{3-} + H^+$$

and several distinct inflexion points may be observed in the titration curve. To observe more than one equivalence point, the equilibrium constants for the various stages (e.g., K_1, K_2, K_3) must differ greatly (e.g., by a factor of 10^4 or more).

In such cases the titration curve may be considered as a combination of three segments; the titration curve for $K_a = K_1$ is joined on to a section appropriate to $K_a = K_2$, which then merges into the titration curve appropriate to a weak acid having dissociation constant K_3.

Tabulations of dissociation constants for most common acids and bases are included in most chemical data books, handbooks, etc.

B. Visual Indicators

Visual indicators may be defined as chemical compounds which facilitate the detection of the equivalence point in titrations by causing some change in the color or opacity of the solution at this point.

An ideal acid-base indicator is a species which responds to the marked change in pH which occurs in the vicinity of equivalence point. It therefore has to be a species which can itself lose or gain protons. In addition, it must possess the property of having a different color in the acid and deprotonated forms. As an acid, its reaction with the solvent is defined by a characteristic acid dissociation constant, i.e.,

$$HIn + H_2O \rightleftarrows H_3O^+ + In^-; \quad K_{In} = \frac{[H^+][In^-]}{[HIn]}$$
(color A) \qquad\qquad\qquad (color B)

If an indicator is added in small amounts to a titration system, the position of equilibrium of the indicator system is controlled mainly by the concentration of the major reactants. Thus the $[H_3O^+]$ in solution may be attributed to an excess of acid or base, and to satisfy the indicator equilibrium equation, the ratio $[In^-]/[HIn]$ has to vary. Since the two species are of different color, one can write

$$\frac{[In^-]}{[HIn]} = \frac{[Color\ B]}{[Color\ A]} = \frac{K_{In}}{[H_3O^+]}$$

In all cases where an indicator is added, both colored forms of the indicator are present in the solution at a given time, but the proportion of each form varies in accordance with the magnitude of K_{In} and $[H_3O^+]$.

The human eye cannot readily distinguish the presence of a second color in solution if one color is present in greater than a 10-fold excess. Thus, if [color A]/[color B] is >10, the eye observes color A; if [color A]/[color B] is < 0.1, the eye observes color B. The excess required can vary from observer to observer and with the colors A and B but the value of 10 is a reasonable approximation.

For a sharp transition in color, there must be a sharp transition in the ratio of the two forms of the indicator. Thus, in a titration conditions must be adjusted so that one drop before equivalence point the color ratio is >10, and one drop later < 0.1, or vice versa.

In a given titration , $[H^+]$ at points near the equivalence point is fixed by the system, and the only variable is K_{In}. In order to be a suitable indicator for a given titration, the K value must be of such magnitude that when substituted in the equilibrium equations, the desired change in ratio occurs between the selected points. If there is a large change in $[H^+]$ in the vicinity of equivalence point, the desired conditions may be met by indicators having a wide range of K values. On the other hand, if the $[H^+]$ change is small, great care is required in selecting a suitable indicator.

The range of pH over which the color observed by the eye changes from A to B, i.e., when the ratio of the indicator forms changes from 10 to 0.1, is known as the color-change interval of the indicator. As shown below, this pH interval is approximately equal to $pK_{In} \pm 1$ (pK = -log K). This approximation may be used if only the value of K_{In} is known, but lists of color-change intervals for various indicators are tabulated in reference books.

If $\dfrac{[In^-]}{[HIn]} = 10$; $\quad \dfrac{K}{[H^+]} - 10$ or $\dfrac{K}{10} = [H^+]$ and pH = pK + 1

If $\dfrac{[In^-]}{[HIn]} = 0.1$; $\quad \dfrac{K}{[H^+]} = 0.1$ or $10K = [H^+]$ and pH = pK -1

The easiest way to assess the suitability of an indicator for a titration is to prepare a titration curve and to superimpose on this curve the color-change intervals of available indicators. The color-change intervals which lie entirely on the near vertical section of the curve represent suitable indicator systems.

C. Sources of Error

Errors in acid-base titrations can arise from

      1. Inaccurate standardization of the titrant

      2. Use of an inappropriate indicator system

      3. Operator difficulty in distinguishing indicator changes

      4. Competing solution equilibria

The strength of any titrant solution should be checked regularly against standard compounds of known composition, purity, and stability. For standardizing acid solutions the compounds recommended include tris(hydroxymethyl) amino methane, $(HOCH_2)_3CNH_2$; mercury(II) oxide, HgO; potassium bicarbonate, $KHCO_3$; sodium carbonate, Na_2CO_3; borax $Na_2B_4O_7 \cdot 10H_2O$. For checking alkaline compounds the recommended primary standards include benzoic acid, C_6H_5COOH and potassium hydrogen o-phthalate, $o-C_6H_4(COOK)$ (COOH).

In titration studies, most analysts prefer to use visual indicators, but it is essential that the suitability of the chosen indicator be checked by experiment, i.e., multiple determinations of the species of interest should be carried out using different selected indicators. The results should then be treated statistically to evaluate the relative merits of the indicators in detecting the true equivalence point.

A statistical approach using several different analysts can also be used to trace errors arising from operator bias.

Continual difficulty in detecting end points or persistent large errors are indicative of competing reactions in the assay solution.

Potential interferents are ions which tend to react with protons, hydroxyl ions, or the indicator anion.

For example, if an acid solution contains iron(III) and aluminium(III) ions, the addition of standard base initially causes the pH to rise due to removal of H^+. Above pH 2, however, base is consumed in producing the hydrous oxides of the metal ions, and the titer and pH changes bear no relationship to the original acidity of the solution. It is also possible for foreign ions in the assay solution to react with the indicator species, effectively removing it, e.g., the indicator may be oxidized or reduced, thus eliminating the traditional color transition. Alternatively, a metal-indicator complex or precipitate may be formed.

Salts present can create buffer situations, for example, the inclusion of a salt of a weak acid (e.g., NaA) will result in uptake of protons from mineral acids (through forming HA). This might raise the pH sufficiently to mask the indicator color-change interval.

Should the solution contain a mixture of acids or bases, it is necessary to know which reactions have gone to completion at the time of indicator color change. For example, consider the titration of a mixture of phosphoric and hydrochloric acids with sodium hydroxide solution. Using methyl orange indicator, a color change is observed when all the hydrochloric acid has been neutralized and a single proton has been removed from the phosphoric acid component (producing $H_2PO_4^-$). However, if phenolphthalein indicator is used, no color change is observed until two protons have been extracted from the polybasic acid (i.e., products are $NaCl + Na_2HPO_4$).

If the mixture is not too complex, and if the species differ markedly in their ability to react with the titrant, the components may be determined individually through suitable choice of indicators. In most cases, however, some separation procedure is desirable.

A major limitation of titrimetric methods is this lack of selectivity; the addition of a base can result in the neutralization of all acids present in solution.

In environmental studies, it is also sometimes necessary to define which acid (or base) factor is most important, e.g., the acid waters pumped out of some coal mines are heavily laden with iron sulfate. The acidity of these waters can be measured in terms of proton activity (i.e., pH), free acid (i.e., H_2SO_4 in equilibrium with the salt), or total acidity (i.e., acidity released on total hydrolysis of the heavy metal salts). Similarly, the alkalinity of soils may be due to the presence of several basic components.

It should also be noted that water is not necessarily the best solvent for titrations, and a wide range of nonaqueous titration procedures are now being used, particularly in studies of amino acids, pharmaceuticals, etc.

III. OXIDATION-REDUCTION TITRATIONS

Oxidation-reduction methods of analysis are probably the most widely used volumetric analytic procedures. A variety of visual indicators and electrometric methods are available for detecting the equivalence point in this type of titration, and suitable standards are available for standardizing all the common titrants. For oxidation purposes, the list of titrants include solutions of chlorine, bromine, and iodine; the salts of Ce(IV), and the oxyanions, OCl^-, BrO_3^-, IO_3^-, ClO_3^-, $Cr_2O_7^{2-}$ and MnO_4^-. A list of typical reducing agents would include the salts of Cr(II), Fe(II), Sn(II), and Ti(III); and compounds such as sodium arsenite, sodium thiosulfate, and hydrazine. In ionic systems redox reactions are generally quite rapid, and reactants can be titrated directly to the equivalence point, but for the determination of most organic compounds it is usually preferable to add an excess of titrant, followed by back titration.

A. Standard Reduction Potentials

To understand the role of the various reagents it is desirable to revise our understanding of oxidation and reduction.

Ions are formed from atoms by the loss or gain of electrons and for each atom there is a spontaneous force which tends to cause donation or acceptance of electrons.

$$A \pm ne \rightleftarrows A^{n-} \text{ or } A^{n+}$$

Conversely, ions tend to lose or gain electrons to yield either the original element or a species of different charge.

The loss of electrons by an atom, molecule, or ion, is termed oxidation; reduction is the gain of electrons by such particles.

In a redox reaction there is a transfer of electrons from one reactant species to another, and the tendency of species to transfer electrons can be measured in electrical units. The potential or electromotive force of an oxidation-reduction system is therefore measured in volts (symbol E).

For the reduction process represented by the equation

$$A^{n+} + ne \rightleftarrows A$$

is has been shown that

$$E = E^o + \frac{RT}{nF} \ln \frac{a_{A^{n+}}}{a_A}$$

In this equation, known as the Nernst equation, E is the potential tending to induce the ion A^{n+} to accept electrons, $a_{A^{n+}}$ is the activity of this species (for dilute solutions and simple calculations, the concentration in gram ions per liter is usually substituted for this term), a_A is the activity of the reduced species. By convention, where this species is in the elemental form it is considered to be in its standard state and to have an activity of unity.

In the factor RT/nF, the symbol R is the universal gas constant, T is the temperature in degrees Kelvin, n represents the number of electrons transferred, and F is the Faraday, the quantity of electricity associated with one mole of electrons (96,493 coulombs).

The symbol E^o is known as the standard reduction potential and is the value of E observed when the logarithmic term equals zero. Thus for the system $Ag^+ + e \rightleftarrows Ag$ at 25^oC, $E^o = 0.800$ V, and $E = 0.800 + 0.059 \log [Ag^+]$.

If the reduction process involves the formation of an ion of lower valency, the activity of this reduced species cannot be taken as unity and the simplified equation for calculation purposes becomes

$$E = E^o + \frac{RT}{nF} \ln \frac{[Ox]}{[Red]}$$

where [Ox] and [Red] represent the equilibrium concentrations of the oxidized and reduced form of the species A.

For the reaction $Fe^{3+} + e \rightleftarrows Fe^{2+}$, $E^o = 0.771$ V at 25^oC, and after converting from natural logarithms, the Nernst equation for this system becomes

$$E = 0.771 + 0.059 \log \frac{[Fe^{3+}]}{[Fe^{2+}]}$$

With more complex electron transfer processes, such as

$$MnO_4^- + 8H^+ + 5e \rightleftarrows Mn^{2+} + 4H_2O$$

the appropriate form of the Nernst equation is

$$E = E^o + \frac{RT}{5F} \ln \frac{[MnO_4^-][H^+]^8}{[Mn^{2+}]}$$

The values of E calculated by substitution in a Nernst equation may be considered to represent a measure of the tendency of electrons to be accepted or donated.

It is not possible to measure such values directly, but one can measure potential differences by setting up an electrochemical cell (as shown in Fig. 4.4). If one half-cell (e.g., system A, Fig. 4.4) has a fixed, known potential then by subtracting this value from the measured emf, one has a value for E for the other half-cell.

The reference half-cell to which the potentials of other half-cell systems are referred is the normal, or standard hydrogen electrode.

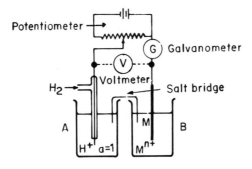

Fig. 4.4 Schematic representation of apparatus suitable for measuring electrode potentials. (Reproduced from Pickering, W.F.: "Modern Analytical Chemistry," Dekker, 1971 with permission of publisher.)

This electrode consists of a platinum wire sealed in a vessel somewhat similar to an inverted test tube. At the end of the wire is a platinum strip coated with platinum black. The tube and platinum strip are immersed in a solution which has a hydrogen ion activity of unity. Hydrogen gas, at unit atmospheric pressure, is admitted to the glass vessel through a side arm and passes over the platinum black before escaping through perforations in the glass wall. Such an electrode has been assigned an arbitrary half-cell potential value of 0.000 V at all temperatures.

In the experimental setup shown schematically in Figure 4.4, vessel A represents a standard hydrogen electrode, vessel B contains a reducible species M^{n+} in contact with its reduced form M. If system B is in its standard state (i.e., the activity of M^{n+} equals unity), then the observed potential (V) between the two electrodes may be recorded as the standard reduction potential for the system M^{n+}/M. The reading of V may be positive or negative, a positive value indicating that the system in vessel B has a greater tendency to accept electrons than H^{+}.

Where the redox couple involves two ions of the same element, e.g., Fe^{3+} and Fe^{2+}, the electrode used in vessel B is usually inert, e.g., platinum. Its role is to promote reversible electron exchange.

For analytical applications, formal potentials are somewhat more meaningful than standard potentials. The formal potential of a system is the potential observed against a standard hydrogen electrode when the reactants and products are at one formal concentration (i.e., one gram formula weight per liter) and the concentrations of any other constituents of the solution are carefully specified. Formal potentials partially compensate for activity effects and errors resulting from side reactions. A formal potential is usually denoted by the symbol $E^{0'}$. A selected list of standard and formal reduction potentials is given in Table 4.1.

It may be observed that formal potentials can differ significantly from the standard potential. This can be attributed in part to activity effects, but very often a major cause is complex formation. This effect may be observed by comparing the formal potential of complex ions with the standard potential of the parent metal ion.

B. Potentiometric Titrations

In setting up a cell for general measurements, the hydrogen electrode is usually replaced by an alternative half-cell combination. The systems chosen (known as reference electrodes) must be capable of

Table 4.1

Selected Examples of Standard and Formal Reduction Potentials[a]

Half-cell reaction	E^O or $E^{O'}$ (volts)
$Ce^{4+} + e \rightleftarrows Ce^{3+}$ (1 F $HClO_4$)	1.70
$Ce^{4+} + e \rightleftarrows Ce^{3+}$ (1 F HNO_3)	1.61
$Ce^{4+} + e \rightleftarrows Ce^{3+}$ (1 F H_2SO_4)	1.44
$MnO_4^- + 4H^+ + 3e \rightleftarrows MnO_2 + 2H_2O$	1.695
$MnO_4^- + 8H^+ + 5e \rightleftarrows Mn^{2+} + 4H_2O$	1.51
$MnO_2 + 4H^+ + 2e \rightleftarrows Mn^{2+} + 2H_2O$	1.23
$Cr_2O_7^{2-} + 14H^+ + 6e \rightleftarrows 2Cr^{3+} + 7H_2O$	1.33
$Br_2(aq) + 2e \rightleftarrows 2Br^-$	1.087
$Br_3^- + 2e \rightleftarrows 3Br^-$	1.05
$Hg^{2+} + 2e \rightleftarrows Hg$	0.854
$Ag^+ + e \rightleftarrows Ag$	0.7995
$Fe^{3+} + e \rightleftarrows Fe^{2+}$	0.771
$Fe^{3+} + e \rightleftarrows Fe^{2+}$ (1 F HCl)	0.70
$Fe^{3+} + e \rightleftarrows Fe^{2+}$ (1 F H_2SO_4)	0.68
$Fe^{3+} + e \rightleftarrows Fe^{2+}$ (0.5 F H_3PO_4, 1 F H_2SO_4)	0.61
$I_2(aq) + 2e \rightleftarrows 2I^-$	0.6197
$I_3^- + 2e \rightleftarrows 3I^-$	0.5355
$Cu^{2+} + 2e \rightleftarrows Cu$	0.337
$Hg_2Cl_{2(s)} + 2K^+ + 2e \rightleftarrows 2Hg + 2KCl(sat)$	0.2415
$AgCl + e \rightleftarrows Ag + Cl^-$	0.2222
$AgBr + e \rightleftarrows Ag + Br^-$	0.073
$AgI + e \rightleftarrows Ag + I^-$	-0.151
$2H^+ + 2e \rightleftarrows H_2$	0.0000
$Pb^{2+} + 2e \rightleftarrows Pb$	-0.126
$Zn^{2+} + 2e \rightleftarrows Zn$	-0.763

Table 4.1 (Continued)

Half-cell reaction	E^o or $E^{o\prime}$ (volts)
$Zn(NH_3)_4^{2+} + 2e \rightleftarrows Zn + 4NH_3$	-1.04
$Cd^{2+} + 2e \rightleftarrows Cd$	-0.403
$Cd(CN)_4^{2-} + 2e \rightleftarrows Cd + 4CN^-$	-1.09
$Na^+ + e \rightleftarrows Na$	-2.714

[a] Based on data from L. Meites, ed.: "Handbook of Analytical Chemistry," McGraw-Hill, New York, 1963.

maintaining a constant known potential during the period of measurement. Typical examples are the silver-silver chloride electrode and the saturated calomel electrode.

With the potential of half-cell A fixed through the use of a reference electrode, any change in the potential of system B causes a change in the potentiometer readings. This principle is utilized in the technique known as potentiometric titration.

The potential of the system in B is approximated by the equation

$$E = E^{o\prime} + \frac{RT}{nF} \ln \frac{[Ox]}{[Red]}$$

and during a titration one changes the ratio [Ox]/[Red]. Consider the titration of a reducing agent (e.g., 0.1 M iron(II) sulfate) with an oxidizing agent (e.g., 0.1 M cerium(IV) sulfate) in say 1 M H_2SO_4. As the titration proceeds, the proportion of the iron salt in the oxidized form continually increases, thus varying the ratio $[Fe^{3+}]/[Fe^{2+}]$. This results in a gradual increase in the observed value of E (as shown in Fig. 4.5) until such time as nearly all reducing agent has been consumed. The actual position of the titration curve segment on the potential scale depends on the magnitude of $E_B^{o\prime}$, e.g., if neither species is specifically subject to complex formation, when 50% of the iron has been oxidized, $[Fe^{3+}] = [Fe^{2+}]$, and $E = E_B^{o\prime}$.

After the equivalence point, the solution contains insignificant concentrations of the reduced form of B (i.e., Fe^{2+}) and the iron system ceases to influence the potential of the solution. However, during the titration, an equivalent amount of the reduced form of the titrant (e.g., Ce^{3+}) is produced, and the presence of excess titrant (e.g., Ce^{4+})

allows the potential of the system to be determined by the oxidant species. Because the oxidant in a reaction possesses a more positive $E^{0\prime}$ value, there is an initial large potential jump which continues to increase gradually as more titrant is added.

The plot of potential (E) versus volume of titrant added is known as a potentiometric titration curve, and as indicated by Figure 4.5, the magnitude of the emf change in the vicinity of the equivalence point is determined by the relative magnitudes of the formal reduction potentials for the reducing agent and oxidant involved.

Hence in the titration of iron(II) salts, a larger jump is observed if the titrant used is cerium(IV) rather than permanganate or dichromate ions.

A large number of potentially useful reactions proceed at a very slow rate and thus are not adaptable to titrimetric procedures. In

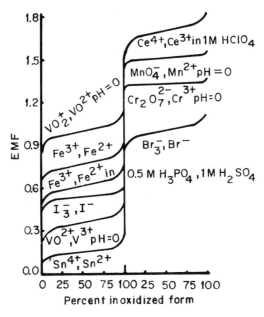

Fig. 4.5 Curves illustrating how the potential of a redox couple varies with the amount present in the oxidized form. Systems on the right-hand side have been plotted in the reverse direction to indicate how combination of data permits construction of titration curves. (Reproduced from Pickering, W. F.: "Modern Analytical Chemistry," Dekker, 1971 with permission of the publisher.)

some cases the rate can be accelerated to an acceptable level by the addition of a catalyst or by increasing the temperature of the reaction, but preliminary experimental evaluation is essential.

As in acid-base studies, the preferred method of ascertaining the form of the titration curve is by experiment. If the assay solution contains several species which can react with the titrant, multiple inflexion points will be observed provided there is a significant difference in the magnitude of the respective formal potentials (e.g., > 0.3 V). Where there is similarity in formal potentials, combined effects are noted. In other words, oxidation reduction titrations are not generally very selective.

One use of a potentiometric titration curve is to aid the selection of an appropriate visual indicator (Section C). More generally it is used directly to determine the equivalence point.

For example, if the reaction involves only ions of variable valency, a platinum electrode responds to the change in the solution potential and a galvanic cell can be formed by inserting a reference half-cell (e.g., Ag/AgCl) and a Pt wire in the assay solution. A potentiometer may then be used to monitor the change in emf as titrant is added. The point of inflexion marks the end-point of the titration.

By using more specific indicator electrodes the scope of potentiometric titrations can be extended beyond simple oxidation-reduction reactions. For example, a silver electrode can be used to observe the changes in $[Ag^+]$ which occur in titrations of species which form sparingly soluble salts with silver ions (e.g., halides, thiocyanate, cyanide ions). Modern glass electrodes develop a potential which is proportional to the hydrogen ion activity in solution (i.e., act like a hydrogen electrode), and measuring cells composed of a glass electrode and a reference half-cell are used to determine the pH ($-\log [H^+]$) of solutions or to monitor pH changes during acid-base titrations.

A modern development, which extends the range of potentiometric titrations, is the use of selective-ion electrodes (Section VI of Chapter 5) to monitor the change in a particular ion concentration during titration.

C. Visual Indicators for Oxidation-Reduction Titrations

In many oxidation-reduction titrations the equivalence point can be detected by means of a visual indicator. The chemical systems used can be divided into two main classes, redox indicators and specific color formers.

A redox indicator is a substance which can accept or donate electrons and whose oxidized and reduced forms differ in color.

For the indicator half-cell system, one can write an appropriate Nernst equation

$$E = E_{In}^{o\prime} + \frac{0.059}{n} \log \frac{[Ox]}{[Red]}$$

If one assumes that the intensity of color is directly proportional to the concentration, and that a tenfold excess is necessary to distinguish the color of a particular form, it can be shown that the color-change interval of the indicator approximates $E_{In}^{o\prime} \pm 0.059/nV$.

Since indicators are added in small amounts, the value of the solution potential E is controlled by the titration systems. The ratio of color forms (i.e., log [Ox]/[Red]) therefore changes during the titration, and a suitable indicator is one whose color change interval lies entirely within the potential jump observed in the vicinity of equivalence point.

Indicators classified as specific color formers can be subdivided into two major subgroups.

  1. Intensely colored titrants whose slightest excess can be detected by its own color. In this group is potassium permanganate, a very popular titrant. It is used, inter alia, for the determinations of iron(II), manganese(II), molybdenum(III), hydrogen peroxide, antimony(II), arsenic(III) [catalyst ICl], oxalic acid [autocatalyzed by Mn(II)], nitrites, and organic compounds.

      Iodine, a weaker oxidant, has proved very useful in determination of sulfide ions, sulfites, thiosulfates, tin(II), and arsenic(III). It can act as its own indicator but the sensitivity of end-point detection is regularly enhanced by the addition of starch, with which it forms an intense blue color.

  2. Auxiliary color intensifiers. The use of starch to detect minute traces of iodine is probably the best-known color intensifier. Another example is the use of thiocyanate ions to detect the first small excess of iron(III) salts.

Where assay solutions are self colored or turbid, or where the difference between the formal potentials for the titrant and titrate system is less than 0.4 V, it is preferable to follow the progress of the redox titration by means of potentiometric measurement techniques.

By varying the titrant, and by using indirect procedures, redox methods can be used to examine any material which contains a species capable of accepting or donating an electron.

Unfortunately, the procedures can be subject to a number of interference effects. For example, if hydrochloric acid is used to dissolve the sample, $KMnO_4$ is an unsuitable titrant because of the tendency of chloride ions to reduce the oxidant (in such cases $K_2Cr_2O_7$ could be substituted). Dissolution with nitric acid introduces a competitive oxidant which may attack the indicator or react with any reducing agents introduced.

A preliminary step in many determinations is preoxidation or pre-reduction of the element of interest, and failure to destroy the excess of any reagent used for this purpose can vitiate the results of any subsequent titration.

IV. ABSORPTION OF RADIATION

Many quantitative determinations are based on observation of the energy transitions which occur when energy is absorbed by the constituent atoms or molecules of a sample.

When reasonably isolated, every elementary system appears to have a number of discrete energy states. To change from one energy state to another, the system must be exposed to heat, radiation, or high-energy particles, the energy supplied being at least equal to the energy difference between the states. Because of the discrete nature of the states, the system can only absorb definite fixed amounts of energy, defined as quanta. On removal of the exciting source, the system reverts to a lower energy state, releasing the excess energy as heat or radiations of characteristic frequency, i.e.,

System + energy (heat, radiation) → Excited state → ground state +
    Energy (radiation, heat)

Electromagnetic radiations possess energy defined by their frequency, and samples can be characterized by studying the nature of the radiations absorbed. The term frequency (ν) arises from the wave nature of radiation; it represents the number of wave crests that pass a given point per second. Frequency is related to energy through Planck's constant ($E = h\nu$) and to wavelength (λ) via the velocity of the radiation ($\nu = c/\lambda$).

Radiations falling within particular ranges of frequencies have been given special names such as x-rays, ultraviolet light, visible light, infrared radiations, etc., and because they possess different energy contents, each range induces a different type of transformation within sample systems.

A. Atomic Absorption

When the system being excited is an atom, energy is absorbed through the promotion of electrons to orbital states of higher energy. The difference in energy between the outer (i.e., higher principal quantum number) orbitals of an atom are generally of the order of 160 to 600 kJ/mol; this corresponds to the energy associated with visible and ultraviolet radiation.

Accordingly, if one succeeds in converting sample components into gaseous atoms, it is possible to have absorption of some characteristic wavelengths of uv - visible light; the amount of a given wavelength absorbed provides a basis for quantitative determinations.

At present, in most of the analytical applications of the technique, atomic vapor is formed by spraying a solution of the sample into a hot gas flame.

When an aerosol enters a flame the following events occur in rapid succession:

1. The solvent is vaporized leaving minute particles of dry salt.

2. The heat of the flame melts or vaporizes the salt, and part or all of the gaseous molecules dissociate to yield atoms.

3. Some atoms absorb sufficient heat to become "excited," and on deactivation emit characteristic radiation [$A + \text{heat} \rightarrow A^* \rightarrow A + h\nu$ (uv-visible light)].

4. Some atoms unite with other atoms or radicals present in the flame.

For most systems the proportion of the atoms that exist in an excited state is very small. The relation between the number of atoms in the ground state N_o and in any particular excited state N_j is given by

$$\frac{N_j}{N_o} = \frac{P_j}{P_o} \exp\left(\frac{-E_j}{kT}\right)$$

where P_j and P_o are the statistical weights of the two states, E_j is their energy difference, k is Boltzmann's constant, and T is the temperature. For the most favored transition involving sodium atoms, N_j/N_0 has values of ca. 10^{-5} at 2000 K, and 10^{-2} at 5000 K; for zinc the corresponding values are 10^{-14} and 10^{-6}.

For the ground state atoms to efficiently absorb radiation (and become excited or higher energy atoms), the energy content of the radiation must correspond exactly to the energy jump associated with the electronic transition (i.e., $h\nu = E_j$).

This stringent requirement can be met by using the characteristic atomic radiation of the element of interest (number 3 above) as the light source in an absorption unit. A hollow cathode lamp in which the required energy is provided by an electrical discharge has proved to be an ideal source of characteristic radiations for absorption studies (Figs. 3.2 and 5.2).

The basic principles of the analytic technique, known as atomic absorption spectroscopy, have been briefly mentioned in a preceding section (Section I.B. of Chapter 3) and are considered again in the next chapter. (Section V.B.). Accordingly, discussion in this section is restricted to comments on the potential errors or interference effects which can be associated with the atom formation process.

Samples can be converted into gaseous atoms by processes other than by introduction into a hot flame. As indicated, the heat of electrical discharges produces vapors which are useful for emission purposes; one can also use this approach for absorption studies. Another alternative approach to atomization utilizes electrical energy to heat a resistance rod containing a small amount of sample. In a limited number of cases (e.g., mercury), chemical reactions are used to reduce the species of interest to an elemental form which has sufficient vapor pressure to be swept by a stream of air into an absorption cell.

In all approaches it is necessary to ensure that the proportion of the sample converted to ground state atoms is constant. Moderate changes in the temperature have little effect but the composition of the plasma environment is extremely important. For example, the interactions which occur in one type of flame can lead to molecular species (e.g., MO or MOH where M is a metal); this not only reduces the observed atom concentration but may produce background interference due to molecular absorption. The radicals present in other types of flames (e.g., C_2H_4 - N_2O) tend to prevent formation of molecular species, and so enhance the observed atom population. Using too high a temperature when elements of low ionization potential are present results in a significant proportion of the atoms being converted into gaseous ions.

Interactions between atoms, and molecular species introduced with the sample, can further reduce the efficiency of atomic absorption. These effects may be minimized by the addition to the assay sample of complexing agents or competing agents. For example, phosphate ions interfere in calcium determinations; this is overcome by adding some strontium or lanthanum salts.

The number of elements which can be determined by this technique is restricted partly by the availability of radiation sources (i.e., hollow cathode lamps) and partly by the ability to produce atomic vapors. With appropriate selection of atomizing technique (e.g., flame type) up to 70 elements can now be examined, with sensitivities falling generally in the μg/ml range. In atomic absorption spectroscopy, sensitivity is defined as "that concentration of an element in aqueous solution which absorbs 1% of the incident radiation passing through a cloud of the atoms being determined." The procedure is also highly specific, the only spectral interference effects being nonatomic absorption (e.g., attributable to scattered light) and emission of radiation in the flame.

A blank correction for nonatomic absorption (or scatter) can be derived by substituting a general band source (e.g., a deuterium lamp) for the hollow cathode lamp and observing any loss in signal. To mask out most of the spectral background, all radiation is passed through a monochromator and only the characteristic frequency (i.e., $h\nu = E_j$) is monitored. To eliminate the effect of atomic emission from the sample, the hollow cathode lamp sources are modulated, and the detection and amplifier unit is tuned to match the oscillation frequency.

Since any radiations emitted are also characteristic of the particular atoms being excited by the source, the intensity can be used as an analytic signal. Measurements are made by placing a detector (photomultiplier tube) at right angles to the original light beam. The technique based on this principle is known as atomic fluorescence spectroscopy.

As noted in later sections, atomic absorption spectroscopy is currently one of the most widely used techniques in trace analysis, but for maximum accuracy it is necessary to choose the most appropriate mode of atom formation, and one must calibrate carefully, e.g., by using standard solutions of similar chemical composition to the assay solutions.

B. Absorption of X-Rays

The energy gap between the inner (i.e., low principal quantum number) orbitals of atoms is of the same order as that possessed by x-rays (ca. 10^5 kJ/mol). Thus if an inner electron is ejected from an atom (through collision with a high energy electron or through absorption of x-rays), and the vacancy is filled by a neighboring electron from another energy orbital, then the excess of energy possessed by this electron is released as characteristic radiation of appropriate frequency (in the x-ray region).

One result of subjecting samples to a beam of high-energy x-rays is accordingly the production of secondary x-rays, the latter having frequencies characteristic of the elements present. Some beam intensity is also lost through scattering of the incident radiation, and while there are some quantitative procedures based on measurement of total x-ray absorption, the majority of applications involve measurement of the frequency and intensity of the secondary x-radiations.

The basic components of the equipment used are shown in Figure 4.6.

Radiations from a high-intensity x-ray tube are directed onto a broad area of sample (polished solid, packed powder, or solution). The secondary emissions are directed on to an analyzing crystal by means of which components are diffracted in accordance with the Bragg equation:

$$n\lambda = 2d \sin \theta$$

where n is an integer, λ is the wavelength of emitted x-radiation, d is the distance between the atomic planes in the crystal, and θ is the angle made by the beam with the diffracting planes. By varying θ, the various wavelengths are brought to focus on the radiation detector, and the intensity noted.

General qualitative analysis of an unknown sample can be performed by automatically scanning the x-ray emission spectrum. By solving the Bragg equation for each peak, the wavelengths of the various radiations may be ascertained and the elements identified through reference to standard tables.

For quantitative analysis, the intensities of one or more of the spectral lines emitted by the element of interest are accurately measured. The measured intensity I_u of a given line of the unknown is then compared with the intensity I_s of the same line emitted from a standard sample of similar composition.

For a given concentration of element, the emitted intensity varies with the nature of the base matrix, because some of the secondary

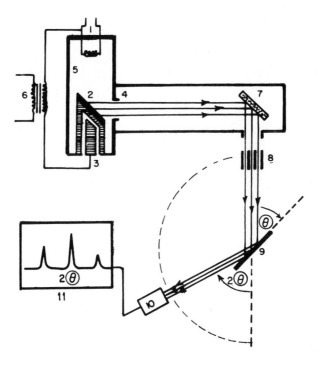

Fig. 4.6 The components of an x-ray emission spectrometer: 1. In-
candescent filament heated by a source of low voltage, 2. metal target,
3. water-cooled support for the target, 4. x-ray tube window, 5. evacuated envelope, 6. high-tension transformer supplying 20 to 100 kV,
7. sample for analysis, 8. collimator slits, 9. flat analyzing crystal,
10. radiation detector unit, and 11. count meter or chart recorder.
(Reproduced from Pickering, W. F.: "Modern Analytical Chemistry,"
Dekker, 1971 with permission of the publisher.)

radiations are reabsorbed by the sample itself. This loss through absorption can be followed by enhanced emission of the lower energy
x-radiations.

Correction for matrix variability can be difficult and computer
programming is regularly used to assist in this problem. Despite this
limitation, x-ray emission spectroscopy has been widely adopted; trace
and macro-amounts of all elements of atomic number > 12 are regularly
determined.

A very specialized and important development is the electron probe microanalyser. In this instrument a finely focused beam of high-energy electrons is used to excite characteristic x-rays from a minute sample area (e.g., 1 μm^2). The emitted x-rays are then separated in terms of wavelength (or energy content) and the intensity of individual emissions determined. The technique permits study of small inclusions, variations in composition across a plane, and examination of the elemental composition of individual particles.

C. Absorption by Molecular Species

In a molecule, the constituent atoms vibrate to and from each other in a characteristic manner and the molecule as a whole can rotate in space. The whole system is thus quantized with respect to electronic energy levels, vibrational frequencies, and rotational energy states.

The total energy of the molecule is given by

$$E_{total} = E_{electronic} + E_{vibrational} + E_{rotational}$$

As indicated schematically in Figure 4.7, the three components of the total energy vary in magnitude by factors of about 10. For example, energy equivalent to about 400 kJ/mol is required to cause an electronic transition, about 40 kJ/mol is needed to cause vibrational changes and changes in rotational states absorb energy measured in kJ per mol.

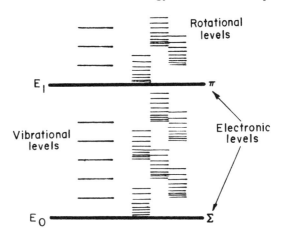

Fig. 4.7 Schematic representation of the energy levels in a molecule. E_0 represents the ground electronic state and E_1 an excited electronic state. Vibrational levels are indicated by broad lines, and rotational levels are shown as finely spaced lines. (Reproduced from Pickering, W. F.: "Modern Analytical Chemistry," Dekker, 1971 with permission of the publisher.)

A molecule is normally in the ground electronic state and the ground vibrational state (at room temperature), but higher rotational levels are involved because only small amounts of energy (e.g., a change in room temperature) are needed to cause changes in the rotational levels. Conversely, the absorption of sufficient energy to cause electronic transitions (i.e., promotion of an electron to a higher energy molecular orbital) results in many changes in vibrational and rotational states.

1. Absorption of Ultraviolet and Visible Radiation

Electronic transitions in molecules can be induced through the absorption of ultraviolet and visible radiation. The energy requirements are quantized, but because of the accompanying vibrational and rotational changes, absorption peaks are quite broad as shown in Figure 4.8.

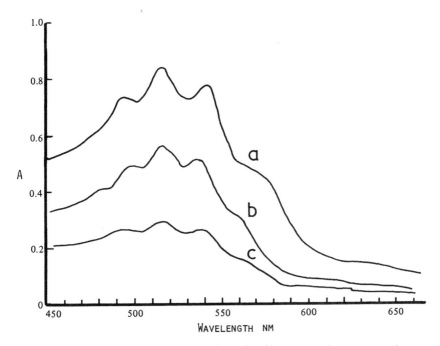

Fig. 4.8 Absorption spectrum of a colored compound. Curves (b) and (c) could represent the spectra obtained on dilution of system (a) or could be unknown samples (i.e., similarity of spectra used for identification purposes.)

With organic compounds absorption is related to a deficiency of electrons in the molecule. Compounds that contain a double bond absorb strongly in the far ultraviolet (e.g., ca. 200 nm) while those containing a conjugated double-bond system (i.e., alternating double and single bonds) absorb at longer wavelengths. The value(s) of the wavelength(s) most strongly absorbed may be used to identify the unsaturated bond groupings (called chromophores) in a sample. The positions of absorption maxima are commonly modified by the presence of various atomic groupings in the molecule. The effect of such substituents is usually greater absorption and a shift of the absorption peak to longer wavelengths.

With inorganic substances strong absorption of radiation is observed if formation of a complex ion creates a stable filled energy level above an energy level not saturated with electrons (e.g., coordination of ligands with transition elements having d orbital vacancies).

If the absorbance of a solution (as determined in a spectrophotometer [Section V.A of Chapter 5]) is plotted as a function of wavelength, the resultant graph is referred to as an absorption spectrum (Figure 4.8). These spectra can be used, on occasions, to confirm the identity of species in solution.

An equally important application is selection of the most appropriate wavelength for quantitative studies. The wavelength of maximum absorption is normally chosen, since this gives high sensitivity and leads to minimum error should there be imprecision in wavelength setting.

Quantitative determinations are based on measurement of the solution absorbance (A) at the selected wavelength, and application of the Beer-Lambert relationship:

$$A = \log \frac{P_o}{P} = \epsilon \cdot b \cdot c$$

[P_o is the intensity of incident radiation; P is the intensity of the transmitted beam; b is the path length; c is the concentration (mol/liter); and ϵ is the molar absorption coefficient, a constant characteristic of the material which varies in magnitude with wavelength.]

The proportionality which exists between the amount of radiation absorbed and the number of moles of absorbing species present can be disturbed by the presence of interfering species or chemical changes involving the absorbing molecules, and is best determined by calibration using a series of standard solutions.

As shown in Figure 4.8, absorption occurs to varying degrees over a wide range of wavelengths, and if the test solution contains several absorbing species, each may make a contribution to the total value noted at some particular wavelength. Correction must be made for this effect through the use of appropriate blanks.

Apparent deviations from the Beer-Lambert law are often attributable to reactions in the solution phase. Solute-solvent interactions, for example, lead to varying fractions of the total content being in a given form (e.g., $Cr_2O_7^{2-} + H_2O \rightleftarrows 2H^+ + 2CrO_4^{2-}$); or the chemical reaction responsible for color formation can be so slow that absorbance varies with time. In these situations control of temperature, pH, nature of solvent, concentration of diverse salts, etc., is necessary if valid quantitative measurements are to be made.

A final factor to be considered is the quality of the measuring instrument, since greater sensitivity and selectivity results from using a radiation source of narrow band width (e.g., wavelength of maximum absorption ± 1 nm).

The absorption of ultraviolet radiation causes some compounds to fluoresce (i.e., emit radiation of longer wavelength, usually in the visible region). This emitted fluorescence may be measured and used for quantitative evaluations.

Related techniques which are applicable to solutions containing suspended particles are nephelometry and turbidimetry. In nephelometry the intensity of the radiation scattered by the particles is measured (Fig. 3.1), while in turbidimetry the quantity measured is the amount of light deflected.

2. Absorption of Infrared Radiation

Infrared spectrophotometers have components analogous to those used in uv-visible units except that the properties of the optical materials necessitate a somewhat different design. (Fig. 3.10.)

The optics of the monochromator, and the sample cells, have to be constructed of material that is transparent to infrared radiation. The bulk of analytic work is done with sodium chloride optics which are transparent over a wavelength range of 1 to 15 μm. The transmitted radiation is usually detected by its heating effects, either by means of a sensitive thermocouple or resistance thermometer.

No solvents are known which do not absorb infrared radiation, although many solvents are transparent over certain ranges of frequency. The choice of solvent for infrared absorption studies is thus limited, and in every case, the measured absorption should be corrected for a solvent blank.

Solid samples can be prepared for analysis by thoroughly mixing weighed portions of powdered sample and finely ground potassium bromide. The mixture is submitted to a pressure of several tons per square inch in a die to produce a highly transparent plate or disk which is inserted into the spectrophotometer. Alternatively, the solid sample is ground with a little heavy paraffin oil to form a mull which is placed between two flat alkali halide disks.

Since one is dealing with vibrational and rotational changes, nearly all chemical compounds absorb infrared radiation and each group of atoms contributes to the overall spectrum. As an approximation, one can consider the vibrational behavior of two atoms connected by a bond to be similar to that of a pair of spheres connected by a spring. For small displacements, the restoring force is proportional to the displacement, and if such a system is set in motion the vibrations are described by the law of simple harmonic motion.

With atomic systems, the rate of vibration is related to the mass of the atoms (M_a, M_b) and the strength of the bond connecting them (represented by a force constant, k).

$$\nu \simeq \frac{[k(M_a + M_b)/M_a M_b]^{\frac{1}{2}}}{2\pi}$$

Should infrared radiation of the same frequency be directed at the molecule, absorption occurs and the amplitude of the molecular vibration increases.

A molecule can be considered to be composed of a number of point masses separated by restoring forces, and changes in the vibration rate of one segment can induce variations in behavior of the molecule as a whole. Thus some of the vibrations detected by ir studies are characteristic of the entire molecule (fingerprint vibrations) while others are associated with certain functional groups. Two major types of vibrations can be identified (1) stretching vibrations in which the distance between two atoms changes without altering the bond axis or bond angles, and (2) bending vibrations. These are characterized by a continuously changing angle between the bonds. Both modes of vibration are quantized.

The recorded absorption spectra can thus be extremely complex but from an examination of a large number of spectra, it has become possible to correlate specific absorption maxima with atomic groupings. Characteristic absorption bands for most of the functional groups present in organic compounds and many inorganic ions have been tabulated. (Table 4.2 contains a small selection, more detailed correlation charts are included in the "Handbook of Chemistry and Physics," monographs on ir spectrometry, etc.)

Table 4.2

Characteristic Infrared Absorption Bands

Vibration	Group	Absorption wavenumber[a] (cm^{-1})
Stretching modes		
C=C	Alkenes nonconjugated	1680–1620
C≡C	Alkyne disubstituted	2260–2190
C—N	Aromatic amines	1340–1250
C=N	Alkyl imines, oximes	1690–1640
C≡N	Alkyl nitriles	2260–2240
	Aryl nitriles	2240–2220
C—O	Alkyl ethers	1150–1060
C=O	Carbonates	1780–1740
	Sat. aliph. aldehydes	1740–1720
	Sat. aryl aldehyde	1715–1695
	Sat. aliphatic carboxylic acids	1725–1700
	Carboxylate anion	1610–1550
		1400–1300
C=S	Organo sulfur compounds	1200–1050
C—H	Alkane (CH$_3$, CH$_2$)	2853–2692
	Alkene–vinyl	3095–3075
	Alkyne	≃ 3300
	Aromatic	≃ 3030
N—H	Secondary amines	3500–3310
	Amine salts	3130–3030
O—H	Alcohols and phenols free OH	3650–3590
	hydrogen bonded	3570–3450
	carboxylic acids	2700–2500

Table 4.2 (Continued)

Vibration	Group	Absorption wavenumber[a] (cm^{-1})
S—H	Sulfur compounds	2600-2550
—N=N—	Azo compounds	1630-1575
Bending modes		
N—H	Primary amines	1650-1590
	Secondary amines	1650-1550
	Amine salts	1600-1575
		$\simeq 1500$
	Primary amides	1620-1590
	Secondary amides	1550-1510
C—H	Alkanes-CH_2	1485-1445
	Alkenes disubstituted	$\simeq 690$
O—H	Carboxylic acids	1320-1210
		950-900

[a] Wave number = (Wavelength)$^{-1}$.

Such data can be used to identify the types of bonding and the groups present in an unknown substance or mixture of substances. The conclusions drawn must be treated with some caution, however, since the frequency at which a particular group absorbs tends to be altered by the presence of other groupings in the molecule. For this reason tabulations indicate regions where one might expect to find peaks corresponding to particular atomic groupings.

The complete infrared spectrum of a pure compound presents a positive method of identification provided that an extensive compilation or atlas of spectra of known compounds is available. It must be emphasized, however, that the sample examined must be extremely pure, since any impurities present will contribute to the overall spectrum.

In principle, quantitative determinations can be made by using techniques similar to those for ultraviolet and visible radiations and by basing calculations on the general relationship $A = \epsilon \cdot b \cdot c$. In practice, difficulties can arise in selecting suitable frequencies for absorption measurements (particularly if a number of compounds is present).

There are problems in preparing samples for examination, and fixing the base line for absorbance measurement is often an empirical process.

The position of the base line can be influenced, inter alia, by neighboring peaks and partial overlaps, and accordingly, it is desirable to scan across the whole peak rather than attempt measurements at a fixed wavelength.

Solutions of the sample are preferred for quantitative measurements, since it is easier to prepare standards (for determination of ϵ) and define the path length b.

With gas samples, analysis for a single component can be performed effectively provided that the desired species has an absorption band which is located in a region where other gases or vapors do not absorb. The effect of pressure has to be considered, but the method is ideally suited for monitoring the composition of gas streams of reasonably simple composition (Fig. 3.3).

V. ELECTROGRAVIMETRY

In oxidation-reduction titrations (Section III) chemical species are used either as sources of electrons or as electron acceptors.

An alternative method of promoting electron transfer is to apply an external voltage to chemically inert electrodes (e.g., platinum) immersed in the test solution.

A. Decomposition Potentials

A simple circuit for electrolytic studies is shown in Figure 4.9. On the imposition of an applied potential, charged ions in solution migrate toward the electrode of opposite sign, and as the potential is gradually increased from zero, an electrical double layer of anions and cations is set up in the vicinity of the electrodes. In maintaining the double layer, and transforming trace amounts of highly reactive impurities in the solvent, a small current (known as the residual current) flows in the circuit [Fig. 4.9b].

Beyond a point known as the decomposition potential, continuous electrolysis occurs and the current increases rapidly with increasing applied voltage. At potentials well beyond the decomposition potential fewer charged species reach the electrodes than can be discharged, the migration of the reacting species becomes the rate-controlling step, and the current flowing in the circuit tends towards a maximum value.

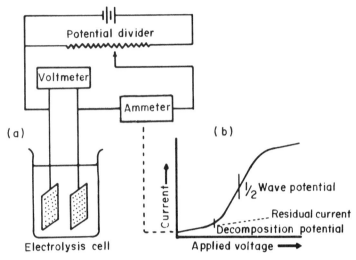

Fig. 4.9 Electrolytic studies: (a) Basic circuit; (b) current-voltage relationship. (Reproduced from Pickering, W. F.: "Modern Analytical Chemistry," Dekker, 1971 with permission of the publisher.)

During electrolysis, oxidation occurs at the anode and reduction occurs at the cathode. The species involved in these electron transfer processes are those which require the least amount of energy for the transformation. For a given species the energy required depends on its concentration, its standard potential, and a variable, known as overvoltage.

After a short period of electrolysis, it can be shown (e.g., by replacing the battery with a wire) that the electrode system is capable of acting as a galvanic cell with a potential equal to $(E_c - E_a)$; E_c and E_a being the potentials of the cathode half-cell and anode half-cell respectively.

Values of E_c and E_a can be calculated for reversible systems by substitution in the appropriate Nernst equation. For example, during the electrolysis of a copper nitrate solution, copper is deposited at the cathode and oxygen is liberated at the anode. The half-cell reactions involved may be written as $Cu^{2+} + 2e \rightleftarrows Cu$; $E^o = 0.34$ V and $4H^+ + O_2 + 4e \rightleftarrows 2H_2O$; $E^o = 1.229$ V. The corresponding Nernst equations are

$$E_c = E^o_{Cu^{2+}, Cu} + \frac{RT}{nF} \ln [Cu^{2+}] \text{ and } E_a = E^o + \frac{RT}{F} \ln [H^+] (P_{O_2})^{0.25}$$

The minimum potential required to cause electrolysis is one which is opposite in direction but greater in magnitude than the electrochemical cell, i.e.,

$$E_{applied} \geq E_a - E_c$$

Comparison of these calculated values with the decomposition potentials E_D observed in experiments often shows a marked discrepancy. Part of the difference can be attributed to the potential between the electrodes arising from the resistance of the solution (E_S = IR) but most of the effect has to be attributed to the influence of irreversible processes occurring at the electrodes. The combined effect of the irreversible processes is known as the overvoltage:

$$E_D = E_a - E_c + IR + \text{overvoltage}$$

Overvoltage effects may arise from irreversible processes occurring at the anode or cathode, and the magnitude is a function of variables such as:

1. The physical state of the product. Metal deposits create small overvoltage effects; gas liberation usually involves high values (e.g., ca. 1.2 V for hydrogen liberation on mercury).

2. The nature, physical state, and surface area of the metal employed for the electrodes.

3. The concentration gradient existing in the immediate vicinity of the electrode. Removal of ions by electrolysis results in the zone near the electrodes having a lower concentration of the reacting species than the bulk solution. In the absence of stirring, the rate of electrolysis becomes dependent on the rate of diffusion of ions across the concentration gradient. (This fact is utilized in polarography, a technique discussed in Section V.C.1 of Chapter 5.) To maintain an appreciable rate of electrolysis, it is thus necessary to increase diffusion rates by increasing the temperature or the applied voltage. Alternatively, one can disperse the concentration gradient by vigorously stirring the solution.

B. Electrogravimetry

In electrogravimetry, the aim is to use exhaustive electrolysis to obtain a quantitative yield of product at one of the electrodes. After deposition the product is washed, dried, and weighed, hence it should be pure, coherent, dense, and smooth. Most determinations involve deposition of a metal on the cathode (e.g., Cu, Ag, Pb, Co, Ni) but a few applications are based on anode deposits (e.g., PbO_2, AgCl).

Factors which help to produce smooth, adherent deposits are efficient stirring, low current densities, and the proper selection of anions. The role of the anions is varied. As a general rule, smoother deposits are produced from solutions of complex ions than from simple salt solutions. Other anions act as depolarizers. For example, the liberation of hydrogen at the cathode at the same time as metal deposition is objectionable (since it produces spongy deposits) and it can be avoided by having nitrate ions in the solution, for these are reduced in preference to hydrogen ions, i.e., $NO_3^- + 10OH^+ + 8e \rightleftarrows NH_4^+ + 3H_2O$

In most electrodeposition studies the current is held more or less constant by periodic or continuous adjustment of the voltage applied to the cell. Initially the system with the most positive reduction potential is reduced at the cathode, but as the concentration of the cation falls through deposition, the rate at which it reaches the electrode decreases. To maintain a constant current the applied voltage has to be increased an and a point can be reached where a second species begins to deposit.

Codeposition is obviously undesirable and may be avoided in one of several ways:

1. Introduction of a depolarizer which is reduced in preference to the second metal to produce a gas or ionic product (NO_3^- in Cu deposition).

2. Chemical removal of the interfering species prior to electrolysis (e.g., selective precipitation).

3. Separation of the deposition potentials of the two species through the addition of a selective complexing agent. (With complex ions, the appropriate value of $[M^{n+}]$ for substitution in the Nernst equation is determined by the stability of the various complexes.)

4. Control of the potential of the cathode, so that it never reaches the value required to begin deposition of the second metal. In this case, the current flowing in the circuit decreases to zero as deposition of the desired species reaches completion.

Electrogravimetric procedures can be extremely accurate, but great care is required if the final deposit is to possess maximum purity. The range of elements which can be determined by this technique is somewhat limited (ca. 15) and most require careful selection of solution conditions.

In pollution evaluation studies, electrogravimetry is rarely applied directly; but as explained in the next chapter some related techniques, such as polarography and anodic stripping voltammetry, have specific advantages.

VI.   COULOMETRY

If experimental conditions can be so arranged that a single reaction of known stoichiometry occurs during electrolysis, then it becomes possible to determine the exact quantity of electricity required to quantitatively reduce (or oxidize) a given species. Faraday's laws of electrolysis state that the amount of chemical change that occurs as a result of electrolysis is directly proportional to the quantity of electricity which passes.

For the general reaction

$$A^{a+} + ne \rightleftarrows A^{(a-n)+}$$

the quantity of electricity required for reduction of one mole of species A is nF coulombs where n is the number of electrons transferred and F is the Faraday (96,487 $\pm$ 1.6 coulombs).

The quantity of electricity (Q) flowing through a circuit is given by the integral of the current flow (i amperes) over the time interval (t seconds), and combination with Faraday's laws yields the relationship:

$$Q = \int_{o}^{t} i \ dt = \frac{nwF}{M}$$

where w is the weight in grams of the species that is consumed or produced during electrolysis and M is its gram molecular weight.

This basic equation indicates that w can be determined directly if Q can be accurately evaluated.

Procedures based on the measurement of the quantity of electricity consumed are termed coulometric methods of analysis. They may be divided into two groups: coulometry at controlled electrode potential, and coulometry at constant current.

Modern instruments electronically monitor the potential and integrate the current flow. The titration cells are designed to give optimum conditions for the electron transfer process. The electrode area is made large, the solution volume is kept small, and a high rate of stirring is maintained. Three electrodes are necessary: a working electrode at which the desired reaction takes place, an auxiliary electrode to complete the electrolysis circuit, and a reference electrode for measurement of the potential of the working electrode.

The experimental conditions chosen are those which give the best compromise between desired accuracy, selectivity, and speed of analysis. In coulometry at constant potential, the substance being determined reacts at an electrode whose potential is maintained at a value that precludes other unwanted electrode reactions. The current decreases exponentially as the electrolysis proceeds and Q has to be evaluated by an integration technique. By varying the cathode potential, successive determinations are possible.

In the technique known as coulometry at controlled current, the electrolysis is performed with a current which is maintained at a constant value. Evaluation of Q thus merely requires the measurement of the current and the time of electrolysis.

With this technique it is more difficult to establish experimental conditions which ensure that a single reaction occurs during electrolysis. As the concentration of the substance being determined falls, a point is reached where migration of this species is insufficient to support all of the current. Some other electrolytic processes must then occur to maintain the current at the selected constant value. However, should the products of the secondary process react rapidly with the substance being determined, the electrolysis can be effectively attributed to a single stoichiometric reaction. The overall result is the same as if a titrant were being generated electrolytically; hence this technique is frequently referred to as coulometric titration.

Consider as an example the reduction of Ce(IV) in a solution containing H_2SO_4 (1 M) and iron(III). On the initial application of a constant current, Ce(IV) is reduced to Ce(III), but as this species is depleted, reduction of iron(III) to iron(II) occurs. The iron(II) is immediately reoxidized by residual cerium(IV) and the efficiency of the electrical reduction process remains 100%.

As in potentiometric titrations, (Section III.B) the detection system commonly used to observe the end of the reaction is a pair of electrodes, one of which is a reference half-cell, the other an indicating electrode. The experimental setup is shown schematically in Figure 4.10. The unit design must ensure that the contents of the cell are mixed rapidly, a and that the indicator system responds quickly. In some arrangements, the titrant is produced in an external generation cell prior to flowing into the test assay.

The generator electrode is usually made of platinum and has an area of 2 to 5 cm^2. The precursor concentration of reagents generally lies between 0.05 and 1 M and generating currents up to 50 mA have been used. Every well-known type of titration, e.g., acid-base, precipitation,

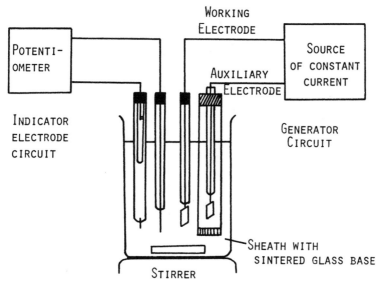

Fig. 4.10 Schematic representation of the components of the reaction cell used for constant current coulometry. (Reproduced from Pickering, W. F.: "Modern Analytical Chemistry," Dekker, 1971 with permission of the publisher.)

complexometric, and redox, has been successfully performed coulometrically, and with coulometric methods it is possible to perform many titrations which cannot be done by classical techniques. Examples are titrations which utilize unstable or difficult to prepare titrants such as bromine, chlorine, chromium(II), copper(I), silver(II), titanium(III), and uranium(IV) or (V) (e.g., SO_2 determination, Section 3.II.B of Chapter 3).

The halogens, Cl_2, Br_2, and I_2, are produced by electrolyzing the appropriate halide salt. Metal ions, such as iron(II), tin(II), and vanadium(IV), are prepared by reducing a compound of higher valency. The ions silver(I), mercury(I), and mercury(II) are generated by using the parent metals as components of the cell anode. A number of reagents are generated by ion exchange, i.e., an ion-exchange membrane charged in the appropriate ionic form is placed in the cell. The reactants (e.g., Cl^-, Br^-, I^-, $EDTA^{2-}$, and Ca^{2+}) are subsequently displaced by competing species liberated by electrolysis, for example H^+ and OH^-. These ions are formed by electrolysis of sodium sulfate or other salt solutions.

The titrations are most conveniently applied to samples containing 10^{-7} to 10^{-5} mole (0.01 to 1 mg) of reactive material. Errors are characteristically of the order of 0.1 to 0.3%. Coulometric methods are ideally suited for both routine and remote analysis, and accordingly, they have been applied to some unusual analytical problems, such as titrations of highly hazardous materials and titrations in molten salt media. Applications include automated analysis and monitoring for specific pollutants (e.g., SO_2).

MONITORING OF WATER QUALITY

I. CRITERIA OF INTEREST

In most cases water quality is monitored to assess the suitability of the water for some subsequent usage. For example, the criteria of quality applied to waters intended to support fish differs from that of water to be used as a public supply.

Table 5.1 summarizes some of the criteria that have been listed for water quality control, and if one adds to these permitted levels of toxicants as contained in clean water acts and other legislative documents, it soon becomes apparent that the range of analyses likely to be involved is extremely broad. In pollution studies, items of interest include the organic content (e.g., biologic oxygen demand, pesticides, etc.), phosphates, heavy metal ions (e.g., Hg, Pb, Cd), and complexing anions such as NTA (nitrilotriacetate), cyanide ions etc.

For regular checking the trend is toward field stations equipped with transmitters for telemetering data to headquarters, or package monitors in which the data are recorded on a magnetic tape. The parameters measured include some or all of the following: pH value, redox potential, chloride ion concentration, dissolved oxygen concentration, conductivity, temperature, and solar radiation. These in situ measurements eliminate much of the uncertainty which can be introduced by changes in composition during transport and storage of samples.

As outlined in Chapter 2, where storage containers are required, they should be carefully cleaned before use, and preferably should be made of pyrex glass, hard rubber, polyethylene or some other inert material. Soft glass containers are only suitable for studies involving constituents which are not affected by storage in such material (e.g., analyses for calcium, magnesium, sulfate, chloride).

Table 5.1

Criteria for Waters
Proposed limits for various uses (in μg/ml)[a]

Species	Drinking water	General farm	Irri- gation	Boiler feed	Chem. industry	Petroleum industry
As	0.05	0.05	1.0	-	-	-
Ag	0.05	0.05	-	-	-	-
Al	-	-	1.0	0.01	-	-
Ba	1.0	1.0	-	-	-	-
B	1.0	-	0.75	-	-	-
Be	-	-	0.5	-	-	-
Ca	-	-	-	0	200	220
Cd	0.01	0.01	0.005	-	-	-
Co	-	-	0.2	-	-	-
Cr(VI)	0.05	0.05	5	-	-	-
Cu	1.0	-	0.2	0.01	-	-
Fe	0.3	-	-	0.01	5	15
Li	-	-	5	-	-	-
Mg	50	-	-	0	100	85
Mn	0.05	-	2	0	2	-
Mo	-	-	0.005	-	-	-
Ni	-	-	0.5	-	-	-
Pb	0.05	0.05	5	-	-	-
Se	0.01	0.01	0.05	-	-	-
V	-	-	10	-	-	-
Zn	5	-	5	-	-	-
SiO_2	-	-	-	0.01	-	50
TDS	500	-	-	0.5	2500	3500

Table 5.2 (Continued)

	Drinking water	General farm	Irri- gation	Boiler feed	Chem. industry	Petroleum industry
Susp. solids	-	-	-	0	10,000	5000
Ca + Mg (Hardness)	60	-	-	0	1000	900

Data selected from Report of Committee on Water Quality Criteria, FWPCA, U.S. Dept. of Interior, 1968.

[a]Limits are regularly updated and proposals vary in different regions. Accordingly, the values quoted should be taken as a guide only.

Changes brought about by adsorption, chemical interaction, or microbiological activity should be minimized by the addition of acid, specific chemicals, or freezing (Table 2.2).

The concentrations of interest (mg to μg/liter) are so low that special precautions are required in the laboratory. Best quality chemical reagents must be employed, and the distilled water used for dilution should preferably be double distilled or subjected to secondary purification with a mixed-bed ion-exchange unit. With each study, one needs to run a reagent blank, that is, distilled water is substituted for the test sample in the analytical procedure. This corrects for trace impurities in the reagents.

The presence of diverse substances in the sample (e.g., chlorine, alum, iron salts, ammonium sulfate, polyphosphates) can interfere with the effectiveness of many analytical procedures, and whenever interference is suspected or encountered, the analyst has to find means of eliminating the interference without adversely affecting the analysis itself.

A qualitative estimate of the presence, or absence, of interfering substances in a particular determination may be made by including some recovery tests. This procedure involves adding known quantities of the substance sought to separate portions of the sample itself. The analytic method is then applied to a reagent blank, a series of known standards, replicates of the test sample, and the recovery tests.

All results are first corrected by subtracting the reagent blank from each of the determined values, and then the standard readings are

plotted graphically. From this graph the amount of sought substance
in the sample alone is determined. This value is then subtracted from
each of the recovery studies. The differences are multiplied by 100
and divided by the amount originally added, to give the percentage recovery. Recoveries differing greatly from 100% reflect either the
presence of interference effects or inadequacy in the method of analysis.

Often the solution is to try a modified procedure or a completely
new analytic technique.

The range of techniques which have been applied to water studies
include titrimetric methods (for acid, base, and redox studies), colorimetric methods, flame photometry, atomic absorption spectroscopy,
emission spectroscopy, polarography, potentiometry (pH and specific
ion electrodes), coulometry, gas chromatography, and automated
analytic instrumentation.

The range of determinations undertaken is equally broad. Accordingly, in this introduction to water monitoring it is not possible to provide comprehensive coverage, but rather it is hoped to give the reader
some appreciation of the approaches available, through discussing
the topic partly in terms of pollutant species and partly through
consideration of a few techniques which are common to a wide number
of determinations, e.g., atomic absorption spectroscopy (used for
metal ion determinations) and ion selective electrodes (used for pH,
anion, and cation determinations).

II. ORGANIC LOADINGS

Organic contaminants (e.g., natural substances and agricultural chemicals) enter water supplies through run off of precipitation, introduction
of sewage and industrial waste, and accidental spills of industrial organic materials.

Some of the contaminants interfere directly with water quality, e.g.,
fish kills are immediate and in such cases a few micrograms per liter
can be significant. In other situations, the organic loading serves
mainly to reduce the concentration of dissolved oxygen, a component
essential for the life of fish and other aquatic organisms. Thus one
can be interested in the determination of the amounts of specific compounds present, or in the total organic content, or both.

For the determination of total organic loadings a number of different types of procedures have been developed. One approach involves
estimation of the reduction capacity (e.g., for oxygen) of the organic
content. Two common procedures in this category are oxygen demand

(biochemical) and oxygen demand (chemical).   Another group of procedures seeks to directly evaluate the organic content.   In this classification one may list determination of the amount of organic material adsorbed by charcoal, evaluation of the total organic content, and individual determinations (e.g., for chlorinated insecticides, nitriles, aromatic ethers, and waste hydrocarbons).

A. Oxygen Demand

There are several approaches to the problem of direct evaluation of the dissolved oxygen content (DO) of waters.

In the chemical method, water samples are treated with manganese sulfate and alkaline potassium iodide solutions containing sodium azide. The dissolved oxygen reacts to produce an equivalent amount of free iodine.   After acidification of the solution the iodine is titrated with standardized sodium thiosulfate solutions.   The procedure is subject to a number of interference effects, and great care is required to avoid trapping atmospheric air in the apparatus during mixing.

For routine monitoring electrical cells are being widely adopted. The units basically consist of two solid metal electrodes in contact with a fixed volume of electrolyte solution.   Reduction of any oxygen present is induced by the application of a suitable potential, and the magnitude of the resultant current flow can be calibrated in terms of oxygen concentration.

In flow system models, the sample is directed between the pair of electrodes and serves as the electrolyte solution.   With constant flow and temperature, the current generated is proportional to the dissolved oxygen content.   To avoid imposing an external emf, the measuring cell sometimes consists of a noble metal cathode (e.g., silver) and a less noble metal anode (e.g., iron-zinc).

In oxygen probes (or sensors) the electrical cell is separated from the sample solution by a membrane permeable to molecular oxygen. For example, in the Beckman polarographic oxygen sensor, a silver anode and a gold cathode are immersed in a gel composed of potassium chloride and a cellulose base.   Contact with the test solution is via a thin Teflon membrane.   This allows oxygen to diffuse into the cell, and on the application of a voltage of about 0.8 V the following electrochemical reactions occur:

Cathode: $O_2 + 2H_2O + 4e \rightarrow 4OH^-$

Anode: $Ag + Cl^- \rightarrow AgCl$

The size of the current flow in the circuit is determined by the rate of diffusion of oxygen to the electrode, which in turn is proportional to the concentration of molecular oxygen in the test solution.

Probes vary in terms of the nature of the electrodes (e.g., some use lead anodes), membrane material (e.g., polyethylene films), and source of applied potential.

Membrane electrodes exhibit a relatively high temperature coefficient (due largely to changes in membrane permeability). They are also sensitive to the salt content of the test waters, and can lose efficiency if the membrane is not cleaned regularly.

Historically, the concentration of organic pollutants in waste waters and effluents has been evaluated by determining the empirical factor known as the biochemical oxygen demand (BOD). This test measures the amount of dissolved oxygen consumed by the sample over a period of five days. To a first approximation, the rate of consumption of oxygen is proportional to the total concentration of the bacteria present, (absorbed oxygen serves as an energy supply for microorganisms). Two major classes of organism are normally present: heterotrophic bacteria which use organic matter both as an energy source and as a source of carbon for growth, and nitrifying bacteria which utilize ammonia or nitrite, principally for energy.

Results are comparative only, since there is no primary standard against which the accuracy of the BOD test can be measured, and unless recommended procedures are closely followed, the presence of nitrogenous species can lead to significant error.

The chemical oxygen demand (COD) of a water is determined by digesting the sample with an oxidant, e.g., acidified dichromate solution, for a period of time, e.g., 2 h, and evaluating the amount of oxidant reduced. The procedure is not specific to organic substances, since it responds to any compound that can be oxidized by this chemical reagent. In addition many organic species (e.g., aromatic hydrocarbons) do not react quantitatively.

B. Carbon Content Studies

Determination of the total organic carbon (TOC) content is more specific, and with modern instruments results can be ascertained in minutes. A number of TOC analyzers use a nondispersive infrared detector to measure the carbon dioxide formed on combustion of the carbonaceous content. A microaliquot of the water sample is injected into a heated, packed tube through which passes a stream of oxygen or purified air. The water is vaporized and the organic content oxidized to carbon dioxide which is measured in a suitable infrared analyzer unit (compare with a CO detector used in air monitoring). Inorganic carbonates interfere; this effect can be minimized by preliminary acidification of the sample.

In other TOC instruments the carbonaceous matter is converted into methane. The organic component of a microsample is first pyrolyzed over an oxidant (e.g., CuO at $850^{\circ}C$) and is then conveyed (in a He/H_2 stream) into a tube packed with nickel catalyst (supported on activated alumina) maintained at 300 to $400^{\circ}C$. The amount of methane produced is determined by means of a flame ionization detector (regularly used in gas chromatography).

The BOD, COD, and TOC values do not always correspond. For example, an input of urea into a water would not affect the BOD and COD values, but TOC would increase. The BOD method indicates the amount of organic matter which is biodegradable during the time and conditions of the test; COD indicates the quantity of material oxidizable by the reagent chosen; and TOC indicates the combined value of bio- and nonbiodegradable organic matter.

Organic loadings can also be compared by determining the quantity of organic material sorbed by an activated carbon bed during passage of large (known) volumes of water (the carbon absorption method, CAM). The bed containing the adsorbed sample is dried and extracted with chloroform. Evaporation of the chloroform leaves a weighable residue of contaminants. This method does not determine the total organic content of water, since the carbon does not adsorb all the organics, and the solvent does not recover all of the materials adsorbed.

Some good correlations between absorbance of ultraviolet light and the concentration of soluble organic matter has been observed in river waters and effluents, and equipment based on this principle is now being developed.

C. Specific Determinations

All the procedures mentioned so far are concerned primarily with bulk effects. Equally important in many systems is identification of individual types of organic residues (e.g., pesticides).

The first obstacle in residue work is the great dilution involved, and it is common practice to extract the organic material from a large volume of water into a small volume of organic solvent. A series of organic solvents may be used successively (e.g., ranging from polar to nonpolar types) but in each case copious quantities of unwanted matter tend to be extracted as well. A separation of the components by a chromatographic process is thus a desirable secondary stage.

Even after chromatographic clean up, the substrate content can far exceed the quantity of residue sought, and hence the determination step must have a high degree of specificity toward the residue. Gas-liquid chromatography is the most popular determinative method, with an electron capture detector being normally used for chlorinated compounds and an alkaline flame detector for organophosphates.

Alternatively, solvent extracts (e.g., using carbon tetrachloride) may be examined in an infrared spectrophotometer to identify functional groups. This technique is poor for identifying mixtures, since the spectrum observed is the sum of all sample components, and as noted previously, preliminary separation by fractionation or chromatography is necessary. With gross contamination of waters (e.g., oil spills) identification can sometimes be made from the initial extract spectrum.

It should be noted, at this stage, that no pollutant is irreversibly residual in a living or otherwise reactive, functioning system.

Many organic species, including pesticides, are very stable but they do not last forever. For example, soil promotes a number of changes and while in cool climates it may take six months for compounds like Dieldrin and DDT to suffer 50% degradation, other pesticides (e.g. heptachlor, malathion, parathion) are completely degraded in a few weeks.

Slow degradation (e.g., DDT) and massive continual dosage into the environment can lead to a build up of the organic species in waters and in the food chain. Thus residue values, such as the European averages shown in Table 5.2, are viewed with some concern.

The total amount A in a system at any time can be expressed in terms of the daily dose (D) and elimination factor (K).

$$A = \frac{[1 - \exp(-Kt)]D}{K}$$

At equilibrium the total is D/K. Since one can determine K by experiment, and estimate the total amount tolerable from toxication studies, it is possible to calculate maximum safe daily doses.

Spillages of polychlorinated biphenyls (used in plastics, paints, rubbers, waxes) create problems because of the toxicity of these materials and their interference effects on analyses for chlorinated residues. Direct analysis is complicated by the fact that there are 210 possible combinations, and commonly a product of this series will have 10 to 20 components.

Table 5.2

Pesticide Residues in Air, Water, and Food[a,b]

Sample	DDT	Dieldrin
Water		
Open sea	<0.001	<0.001
Coastal	0.005	0.007
Ground	0.02	0.03
Rain	0.08	0.04
Air	0.01	0.02
Foods		
Fruit	10	2
Cereals	20	2
Meat	50	10
Fat	200	20

[a] Taken from I. S. Taylor, Proc. Roy. Aus. Chem. Inst., 39:350, 1972.

[b] Average values, expressed as $\mu g/ml$.

Organophosphates or carbonates are more easily degradated than chlorinated pesticides and so are not found as residues in the wider environment, although they sometimes occur due to accidental spillages.

Increases in organic content, if associated with increased concentrations of inorganic plant nutrient species (e.g., ammonia, nitrate, phosphate, silica), can so promote algal growth that bodies of potable or viable water become overgrown with organisms. The resultant increase in biological oxygen demand makes the water useless to aquatic animals (which usually leave) and aquatic plants (which cannot leave and are destroyed). This process is known as eutrophication.

III. PHOSPHATES

One of the many factors to be taken into account in considering the possibility of reclaiming water for industrial or potable use is the phosphate content. Phosphates can interfere in lime-softening processes, or aid hard scale formation in boilers, and under certain conditions contribute to eutrophication of enclosed bodies of water.

The phosphate content can arise from industrial effluents (e.g., metal finishing processes) or from domestic sewage. Phosphates are a normal constituent of human excreta and not all is removed during

sewage treatment.   Other sources are soaps, detergents, and
fertilizers.

The phosphorus can be present in several forms, the main classifications being orthophosphate, condensed phosphates (pyro-, meta-,
and polyphosphates), and organically bound phosphate.

Orthophosphates applied to residential or agricultural land (as
fertilizer) are carried into surface waters with storm runoff.   Condensed phosphates are added in small amounts in some water treatment
processes, but the main source of these species has been heavy-duty
washing powders which contain compounds like sodium tripolyphosphate
($Na_5P_3O_{10}$) as a major ingredient.   Polyphosphates are partially
hydrolyzed to orthophosphate during passage down sewers.   Organic
phosphates are formed primarily in biological processes.

For interpretative purposes, the phosphate content is sometimes
divided (analytically) into groupings representative of these several
forms.

Phosphate analyses embody two general procedural steps: (1)
conversion of the phosphorus form of interest to soluble orthophosphate,
and (2) colorimetric determination of soluble orthophosphate.

The orthophosphate content is usually determined by treating an
acidified solution with molybdate ions (to form a heteropoly phosphomo-
lybdate) and a reducing agent (e.g., ascorbic acid, stannous chloride).
The intensity of the resultant blue coloration is compared with standards
similarly treated, by means of an appropriate type of spectrophotometer
(Fig. 3.9, and Section IV.C.1 of Chapter 4).

To separate filtrable (or dissolved) from particulate forms, filtration through a 0.45-μm membrane filter is widely adopted.

The component in these fractions which responds to the orthophosphate colorimetric test without preliminary hydrolysis or oxidative
digestion of the sample, is reported as the orthophosphate concentration.

Acid hydrolysis at boiling water temperature converts filtrable and
particulate condensed phosphates into orthophosphate, and this is the
first step in the determination of polyphosphate content.   By judicious
selection of acid strength, hydrolysis time, and temperature, release
of phosphorus from organic compounds is kept to a minimum.

The fraction of the total phosphate which is converted to orthophosphate only by oxidative destruction of the organic matter present is
termed the organically bound component.   The severity of the oxidation
required depends upon the form, and to some extent the amount, of
the organic phosphate present.

Dissolved orthophosphate measurements are sometimes described as the reactive phosphorus or inorganic phosphate component. The assay values include labile organic phosphorus species as well as any soluble orthophosphate ions. Subtraction of this value from total phosphate contents provides a measure of the unreactive phosphorus (organic and inorganic) content.

Total phosphate values are obtained by oxidizing the sample (e.g., by boiling with potassium persulphate or fuming with perchloric acid) before proceeding to form the reduced heterophosphomolybdate colored species.

There are several recommended modifications for each basic analytic procedure. It has been demonstrated in interlaboratory studies that great care is required in all to achieve reproducible results, particularly in the concentration ranges commonly encountered in natural waters (orthophosphate levels in surface waters are frequently < 10 μg/liter; total phosphorus in sea water can be as high as 75 μg/liter).

Because of its probable role in eutrophication, it is now considered ecologically desirable to minimize the phosphorus contents of waters, and efforts are being made to increase the efficiency of phosphorus removal in sewage treatment processes. Efforts are also being made to replace phosphates in detergents. The best known substitute is sodium nitrilotriacetate (NTA), but since this is one of the complexones (i.e., similar to EDTA) it readily reacts with heavy metal ions and introduces new problems, such as transmission of toxic metal through treatment plants (as complexes) and disturbance of metal nutrient concentrations.

IV. NITROGEN COMPOUNDS

The total nitrogen content of water samples can be present in many chemical forms, and as with phosphorus, for interpretative purposes it is desirable to ascertain the contribution of each form.

Free ammonia (i.e., ammonium salts) almost invariably originates from animal wastes, and the amount detected can range from 0.02 mg/liter (surface water) to 100 mg/liter (in sewage discharge).

This ammonia content is determined by distillation of freshly collected samples, after the addition of a phosphate buffer to give a pH of about 7.4. The ammonia passes over in the first portion of the distillate, and is quantitatively absorbed in either 0.01 M sulfuric acid or a boric acid solution. The isolated ammonia can be evaluated colorimetrically or by titration.

The colorimetric method most widely applied utilizes the yellow or brown coloration produced on the addition of Nessler reagent.

$$NH_4Cl + [2K_4(HgI_4) + 4KOH] \rightarrow Hg_2NI, \ H_2O + 7KI + KCl + 3H_2O$$

$$\text{Nessler reagent} \qquad \text{Yellow}$$

The sample color intensity is matched with the color produced by standards containing known amounts of ammonium chloride. Provided appropriate steps are taken to minimize interference effects, the Nessler method can be applied directly to some test samples, i.e., without prior distillation.

For titrimetric determination, the ammonia is collected in a 2% boric acid solution containing an indicator which changes color at $pH \simeq 5$ (e.g., methyl red plus methylene blue). The entrapped base is then titrated with standard 0.01 M sulfuric acid until the indicator changes color.

Errors can be quite significant, particularly when low concentrations are involved.

Nitrogenous organic matter present in water samples is converted (wholly or partially) into ammonia when an alkaline solution, to which potassium permanganate has been added, is boiled. The ammonia thus generated is termed albuminoid ammonia. The fraction of the total nitrogen evolved as ammonia varies with the nature of the organic substrate, and determination of this component is now essentially obsolete.

To determine the total nitrogen content, effluents and other liquid samples are acidified with sulfuric acid and evaporated to a small bulk. The residue is then treated by the Kjeldahl process, that is, it is digested with concentrated sulfuric acid containing potassium sulfate (to elevate the temperature) until the solution is clear [a small amount of catalyst, (e.g., mercury salt) is also desirable]. In the digestion the amino nitrogen of most organic materials is converted to ammonium bisulfate.

The acid solution is cooled, neutralized, and any mercury ammonium complex in the digestate decomposed by the addition of sodium thiosulfate. The solution is then made alkaline and the ammonia released is distilled over into a boric acid–absorbing solution. The ammonia content of the distillate is determined as outlined above.

When nitrogenous matter is oxidized by the environment, the nitrogen remains chiefly in the form of nitrate; but nitrite is also occasionally present. The presence of appreciable quantities of nitrite can be regarded as an indication of recent sewage contamination.

Nitrite is readily oxidized to nitrate, and accordingly, large amounts of nitrate points to past sewage contamination.

The procedures used for the determination of nitrate, nitrite, and urea nitrogen contents are generally based on the formation of a colored complex species, the intensity of the color being related to concentration through comparison with standards.

For the determination of nitrate content a number of color-forming reagents have been recommended (e.g., phenoldisulfonic acid, brucine, chromotropic acid). Each method overcomes certain interference effects, but remains subject to others. Thus the analyst must select the method most suitable for the kind of samples being tested. The range of choices can be increased by reducing the nitrate to nitrite (under controlled conditions) with reducing agents such as amalgamated zinc or cadmium.

The color-forming procedure most widely used to estimate the nitrite concentration of a solution is that based on azo dye formation, e.g., at pH 2, in the presence of nitrite, sulfanilic acid is diazotized, and when coupled with a base such as naphthylamine yields an intense reddish purple coloration (compare with determination of oxides of nitrogen, Section III.B of Chapter 3).

The more modern approach to the monitoring of nitrates (and ammonium ions) is to use specific-ion electrodes. This technique is considered in Section IV.

The substitution of nitrilotriacetate (NTA) for polyphosphate in detergents has produced a need for suitable analytical methods for this reagent. It tends to occur in water in the form of metal complexes. The metals can be removed from the sample, however, by reaction with a chelating resin. The freed NTA is then caused to react with a colored complex of zinc (e.g., involving zincon reagent). This gives a decrease in color intensity which is proportional to the concentration of NTA. If other complexing agents (e.g., cyanide ions) are present, it is necessary to use a more specific method such as gas chromatography.

It is obvious (from the summaries quoted in Sections III and IV) that colorimetric procedures are regularly applied in phosphorus and nitrogen studies. Other applications of the technique in water monitoring include the evaluation of chlorine, ozone, phenols, silica, surfactants, and an extensive group of metal ions. Because of this widespread use, the principles of absorption spectrophotometry are considered in some detail in the next section.

V. DETERMINATION OF TOXIC METAL IONS

The metal ion content of waters can be derived from natural processes (e.g., leaching from ore bodies) but most of the values of interest in pollution studies are the result of human operations (e.g., effluent discharge from industry, oxidation and leaching of mine dumps, corrosion of metal surfaces, etc.) The concentrations involved vary with the source of the sample (e.g., direct drainage from plants, rivers, oceans) and the detailed analytical procedures have to allow for differences in base composition (e.g., high salt content of sea samples, acidity of some discharges, presence of organic matter, etc.)

Most procedures, however, can be classified into one of the following categories: spectrophotometric methods, atomic absorption methods, or electroanalytical procedures.

Each of these techniques possesses particular advantages and disadvantages, and it seems most appropriate to consider the evaluation of trace amounts of metal ions in terms of these categories.

A. Absorption Spectrophotometry or Absorptimetry

Absorptimetric analysis is based on the absorption, by samples, of radiation having wavelengths or frequencies which fall in the visible and ultraviolet region of the electromagnetic spectrum. Procedures which compare the intensity of visible colors are also sometimes called colorimetric methods.

The equipment required is some form of spectrophotometer and the basic components of this type of instrument are shown in Figure 5.1 (compare with Fig. 3.9). As noted earlier (Section IV.C.1 of Chapter 4), the unit may be used to identify the most appropriate wavelength for quantitative studies (usually λ_{max} on the absorption spectrum) and routinely provides the absorbance readings (at λ_{max}) which are related to concentration through the Beer-Lambert relationship [$A = \log P_0/P = \epsilon \cdot b \cdot c$].

In a spectrophotometer, the initial source of radiant power is a lamp which emits a broad spectrum of radiation. Selected segments of this radiation are isolated by means of a colored filter or a monochromator (i.e., a dispersing prism or grating with an optical system to focus selected wavelengths on to the exit slit).

The exit opening from the source unit is a slit of variable dimensions. This allows control of the intensity of radiation (P_0) entering the sample cell. The sample cell has to be transparent to the incident radiation (glass for visible studies, fused quartz for ultraviolet work) and usually has lateral dimensions of 0.5 to 4 cm.

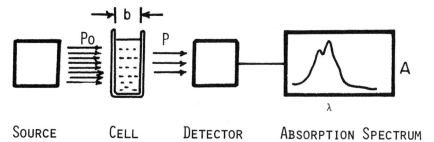

SOURCE   CELL   DETECTOR   ABSORPTION SPECTRUM

Fig. 5.1 Schematic representation of absorption of uv-visible radiation. P_0 is the radiant power (of selected wavelength, λ) entering the sample cell (of path length b cm). P is the radiant power leaving the cell. Plot of absorbance ($A = \log P_0/P$) versus wavelength yields an absorption spectrum.

The intensity of light transmitted by the sample in the cell (P), is measured by a photoelectric detector with associated amplifier and recording circuit.

The molar absorption coefficients (ϵ) of many self-colored species fall in the 10- to 100-liter $mol^{-1} cm^{-1}$ range; hence, in order to obtain absorbance readings in the preferred region (0.2 to 0.7), one requires concentrations (c) to be of the order of 10^{-2} to 10^{-3} M when using cell lengths (b) of a few centimeters. By the same reasoning, when studying say permanganate solutions (ϵ ca. 5000), sample concentrations should be around 10^{-4} M.

In water monitoring, and other trace analyses, the concentrations are normally measured in units of mg/liter (i.e., 10^{-4} to 10^{-5} M), or μg/liter, and if it necessary to extend absorptimetry to these levels, it is necessary to convert the specified solute species into complexes having molar absorption coefficients $> 10^4$.

This color-forming stage, in which a suitable reagent (usually organic) is added to the test solution, is a chemical process and as such is subject to the influence of competing equilibria.

The primary problem is to transform all of the desired constituent (A), or something chemically equivalent to it, into the colored state. If the chromogenic reagent is represented by C, the equilibrium constant for the reaction

$$aA + cC \rightleftarrows A_a C_c \text{ (colored)}$$

should have a large numerical value to ensure that $[A_a C_c]/[A] > 100$.

Since the color reaction is an equilibrium process, the magnitude of the term $[A_aC_c]$ (i.e., the color intensity) will vary with the initial concentration of A (i.e., $[A]_0$) and any other factor that can influence the magnitude of [A] or [C]. For example, in the presence of another ligand L, there can be competition for A (e.g., $A + L \rightleftarrows AL$) which results in either partial or minimal conversion to A_aC_c. The final position is determined by the respective concentrations of the competing ligands and the relative stabilities of the complex species formed.

This effect can be used to advantage when several components of the solution are capable of reacting with the chromogenic reagent. By adding a competing ligand system which preferentially reacts with the interfering species, reaction of the color-forming reagent can be restricted to the one desired component. This process is known as masking.

Of major importance is the pH of the solution, since complex formation quite regularly involves displacement of a proton from the reagent and introduces a $[H^+]^x$ term in the equilibrium equation. On other occasions, other constituents of the sample form stable complexes (not necessarily colored) which consume most of the reagent and leave little for interaction with the desired species A.

If more than one species capable of absorbing radiation is present, the total absorbance measured is the sum of the absorbance due to the individual components, that is,

Total absorbance $A_T = \Sigma_i A_i = b \Sigma_i \epsilon_i c_i$

This additivity can be both a problem (e.g., nonlinear calibration) and useful (e.g., it permits subtraction of the absorbance due to the solvent or reagents or an interfering species). For example, one can estimate the nitrate content of a water by measuring the absorbance of radiation of wavelength 220 nm. Dissolved organic matter also absorbs radiation of this wavelength and so introduces an error. However, unlike the organic matter, nitrate ions do not absorb using radiation of wavelength 275 nm, and one can correct the nitrate value by making a second measurement at this wavelength.

The addition of masking agents, control of pH, and extraction of the colored species A_aC_c into a nonaqueous solvent are some of the means used, either singularly or in conjunction, to ensure that the final solution examined in the spectrophotometer is a quantitative representation of the amount of A present in the initial sample.

Table 5.3 shows how the relevant information is summarized in compilations such as the Handbook of Analytical Chemistry or monographs on absorption spectrophotometry.

Table 5.3

Summary of Procedures for Spectrophotometric
Determination of Cobalt[a]

Reagent and conditions	Wavelength (nm)	Range (mg/liter)	Interferences
Nitroso-R salt; 0.5% in water, pH 5.5 $\pm$ 0.5; Co:R=1:3; destroy excess with 3% $KBrO_3$	425	0.1-1.0	Fe(III); Cl^-, F^-
2-Nitroso-1-naphthol; 1% in glacial HOAc, Cit. buffer; pH 3-4, ext. with $CHCl_3$, wash with NaOH to remove excess	530	0.2-4	Fe(II), Pd, Sn(II)
SCN^-; 44% aq. NH_4SCN, pH 3-5, ext. with i-$C_5H_{11}OH$	312	0.2-10	Bi, Cu, Fe(III), Ni, Ti, U(VI)
2,2′, 2″-Terpyridine; 0.1% in water, pH 2-10; Co:R=1:2	505	0.5-50	Cr, Cu, Fe(III), Ni, V(V)

[a] Based on data from Meites, L., ed.: "Handbook of Analytical Chemistry," McGraw-Hill, N.Y., 1963.

Because it is not possible to ensure that all factors are under control, it is necessary to relate the absorbance of test solutions to concentration by means of a calibration curve.

The calibration graph should be prepared by measuring the absorbance of known concentrations of the test species when present in base solutions which are chemically similar to those of the samples. One procedure for achieving chemical similarity involves the addition of known amounts of standard to aliquots of the test solution. This is known as spiking or internal calibration.

Ideally the calibration plot should be a straight line (of slope ϵb) but deviations are regularly observed. These deviations may arise from instrumental causes (e.g., broad bands of light), or the position of equilibrium may change with dilution, or other absorbing species may be present, etc.

With appropriate care in reagent selection, blank correction, calibration, and sample preparation, the accuracy and precision of determinations based on absorptimetry can be of the order of 2% of content, but the actual value achieved depends on a range of factors, including the concentration levels involved, the nature of the species being determined, the quality of the measuring instrument, etc. For example, it can be shown that the relative percentage error in the photoelectric measurement increases from about $1\frac{1}{2}\%$ to $>8\%$ if one operates outside the recommended absorbance reading range of 0.2 to 0.8.

B. Atomic Absorption Spectroscopy

Chemical analysis by atomic absorption spectrometry involves converting part of the sample into an atomic vapor and measuring the absorption, by this vapor, of radiation which is characteristic of some particular element.

As noted in a previous chapter, in most analytical applications of this technique, the atomic vapor is formed by spraying solutions into a flame. The flame volatilizes the solvent and ultimately causes dissociation of the minute, solid solute particles into atoms.

A small fraction of these atoms absorb additional heat energy by promoting electrons to a higher energy atomic orbital (i.e., they are excited by the flame) and on reverting to the initial (or ground) state the excess energy is released as characteristic radiation. The majority of atoms remain in the ground state, unless they are exposed to radiation whose energy content $(h\nu)$ corresponds exactly to the energy jump associated with the electronic transition. The radiation sorption process is very selective and the method is quite sensitive (ppm range).

$$A + heat \rightleftarrows A^* \xrightleftharpoons[absorption]{emission} A + h\nu$$

The stringent requirement in regard to the energy content of the incident radiation is most readily met by using the characteristic atomic emissions of the same element. The source chosen must emit a spectrum in which the frequencies are well defined, the intensity must be constant, and background emission minimal. A hollow cathode electrical discharge lamp (shown schematically in Fig. 5.2) fed by a stabilized power unit meets these specifications.

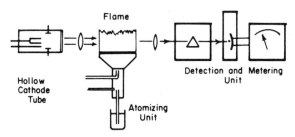

Fig. 5.2 Schematic representation of the basic components of an
atomic absorption spectrophotometer. (Reproduced from Pickering,
W. F.: "Modern Analytical Chemistry," Dekker, 1971 with permission
of the publisher.)

Of the large number of emissions derived from any source, only
a small number show any appreciable absorption by atomic vapor. Accordingly, a monochromator is usually used to isolate the radiation to
be measured from all other wavelengths in the spectrum of the light
source.

The basic components of an atomic absorption spectrophotometer
are therefore a hollow cathode lamp, a flame-atomizer unit, a monochromator, a photoelectric detector, an ac amplifier and rectifier, and
an output meter. The units are shown in diagrammatic form in Figure
5.2.

The flame can be considered to play the same role as the sample
cell in other forms of absorption spectrophotometry. Absorbance of
the characteristic radiation on passage through the flame thus varies
with the length of the absorption path (commonly 5 to 10 cm) and with
the concentration of absorbing atoms present. The latter is invariably
a function of the atomizer unit, the flame type, and the concentration
of the material being sprayed into the flame.

The flame conditions required to produce atoms of different elements
vary. For each element there is a preferred combustion system and
an optimum fuel to oxidant ratio. Flame types which have proved useful include air-coal gas, air-propane, air-hydrogen, air-acetylene,
oxy-acetylene, and nitrous oxide-acetylene flames.

The intensity of the radiation isolated from the source is usually
weak, and the most suitable detector is a photomultiplier tube connected
via appropriate amplifiers to a recording meter.

In single-beam instruments, the recording meter is adjusted to show
full-scale deflection when only the solvent is being aspirated into the

flame. The sample solution is then introduced into the flame via a nebulizing unit which converts it into a fine spray. The change in meter reading observed is correlated with the concentration of the element present by means of a calibration graph. For maximum accuracy the power supply to the lamp and detection systems must be highly stabilized, and the standards used must be chemically similar to the sample.

An important modification substitutes for the flame unit a carbon rod or carbon furnace attachment. Microliter samples of liquid are placed in a small cavity in the carbon sample holder which is then heated by the application of an electrical current. With careful electrical control one can first evaporate excess solvent, and subsequently burn off organic matter before applying sufficient heat to convert metallic residues into atoms. The burst of metal vapors causes a sharp absorption peak on a recorder.

The approximate detection limits for a number of elements, using one commercial instrument, are shown in Table 5.4.

While these levels are very low, the concentrations present in water samples can be smaller, and often preconcentration is desirable. Concentration techniques used include evaporation, chelation and solvent extraction, ion exchange and co-precipitation techniques.

Atomic absorption spectroscopy is reasonably selective, but interference effects do occur and it is advisable to adhere to standardized conditions. The number of potential applications may be gauged from the fact that one operational manual (Water Analysis by AA, Varian-Techtron, 1972) lists direct methods for 24 metal ions (Ag, Al, As, B, Ba, Be, Cd, Ca, Co, Cr, Cu, Fe, Hg, K, Mg, Mn, Mo, Na, Ni, Pb, Se, Si, Tl, and Zn) and three indirect methods (chloride; phosphate and silicate; and sulfate).

C. Electroanalytical Procedures

Colorimetric analysis of metal-bearing wastes and receiving streams is generally difficult and time consuming. Accordingly, it is currently considered preferable to use atomic absorption spectroscopy or polarography. The advantage of polarography lies in the fact that a number of of metals (e.g., Cd, Cu, Pb, Ni, and Zn) may be determined simultaneously.

The basic components of the apparatus required for polarographic measurements are pictured in Figure 5.3.

Table 5.4

Approximate Atomic Absorption Spectroscopy
Detection Limits[a],[b] (μg/liter)

Flame		Carbon Rod[c]	
Limit	Element	Limit	Element
0.3	Mg, Na	0.02	Cd, Mg, Na, Zn
0.6	Ca, Cd	0.05	Ag, Ca
2	Be, Cu, Li, Zn	0.2	Be, K, Mn
3	Ag, K, Mn, Rb	1	Bi, Co, Cr, Cu, Fe, Li, Pb, Sr
5	Cr, Fe, Sn	2	Au, Ni
8	Co, Ni	6	Al, Cs, Sb
10	Au, Ba, Te, Ti	10	Mo, Si, Sn
20	Cs, Pb, Pd	20	As, Hg, Se, V
40	Al, Mo, Si, Sr	40	Pd, Pt
70	Bi, Sb	600	B
100	Pt, V		
250	As, Hg, Ru		
2000	B, Hf, Nb, Ta, W, Zr		

[a] Based on use of appropriate flame conditions and a particular brand of instrument.

[b] Using 5-μl samples.

[c] Detection limit is defined as "that concentration in solution of an element which can be detected with a 95% certainty. This is that quantity of the element that gives a reading equal to twice the standard deviation of a series of at least ten determinations at or near blank level."

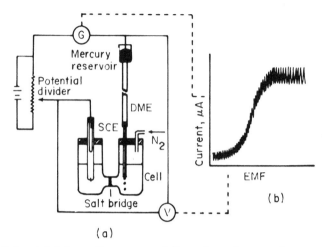

(a)

Fig. 5.3   (a) Schematic representation of a polarographic circuit;
(b), a diffusion current-voltage curve. SCE = saturated calomel electrode; DME = dropping mercury electrode. (Reproduced from
Pickering, W. F.: "Modern Analytical Chemistry," Dekker, 1971 with
permission of the publisher.)

In polarography, the electrolysis which is induced in a cell containing a microelectrode is observed by simultaneously measuring the
current flowing in the circuit and the applied potential. The current-
voltage relationship is then used to determine the identity or concentration of the species reacting at the microelectrode.

The small electrode, at which deposition of metals takes place,
should not alter in nature during analysis, and in practice, the most
useful electrode has been found to be the dropping mercury electrode.
This is composed of a fine capillary tube through which mercury drops
slowly from an elevated reservoir. The other electrode in the circuit
should preferably have a large surface area and a more or less standard
potential. A calomel electrode is commonly used, but for many applications, a pool of mercury has proved satisfactory.

A changing potential of 0 to 3 V is applied across the electrodes
and the current flowing in the circuit, as a function of potential, is
measured with a sensitive galvanometer, or is amplified and fed to a
recorder unit.

The current versus potential plots are known as polarograms and
under ideal conditions are of the form indicated in Figures 5.3 and 5.4.
Multiple curves (such as shown in Fig. 5.4) can be observed when more

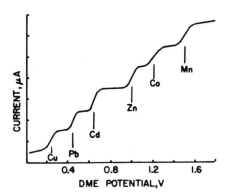

Fig. 5.4   Form of polarogram sought when studying solutions contain-
ing several reducible ions.   (Base electrolyte in this sample 0.1 M
KCl).   (Reproduced from Pickering, W. F.: "Modern Analytical
Chemistry," Dekker, 1971 with permission of the publisher.)

than one reducible species is present in solution.   The position of the
half-wave potentials (i.e., inflection point of curves) is indicative of
the type of species being reduced.

The magnitude of the current jump is proportional to concentration
if conditions are controlled so that reducible ions reach the electrode
solely by a diffusion process.

Diffusion control of the current requires that there be no stirring
and the movement of reducible ions due to migration under the influence
of the applied field has to be minimized by adding to the solution a large
excess of electrolyte whose ions have a decomposition potential which
is much greater than that of the species being examined.

The maximum, or limiting, current recorded includes a non-fara-
daic contribution called the residual current.   Correction for this can
be made by observing the current that flows in an electrolyte solution
that contains all the original chemical compounds except the reducible
species.   For many practical purposes, a reasonable correction is
obtained by extrapolating the first leg of the curve (compare Fig. 5.3b).

The curves obtained in practice are rarely as symmetric as shown
in Figure 5.4, and the shape can be further distorted by the presence of
peaks called maxima.   This effect can be minimized by the addition of
a small amount of surface-active agent to the solution.

The position of the half-wave potential ($E_{\frac{1}{2}}$) for a given element on
the current-potential curves can be altered by the addition of a

complexing agent. For example, the half-wave potential of the $Cu(H_2O)_4^{2+}$ ion is different from the values observed for $Cu(NH_3)_4^{2+}$ and $Cu(CN)_4^{2-}$. The more stable the complex, the more negative its $E_{\frac{1}{2}}$ value.

If a solution contains several reducible species, two or more may decompose at a similar potential, leading to a single distorted peak on the polarogram. This undesirable effect can often be overcome by adding a complexing agent which reacts with one of the species to form a compound of such stability that the half-wave potential is markedly altered.

The electrolyte solution (known as the base solution) added to the polarographic cell therefore contains (1) an aliquot of the sample containing small amounts of reducible species (10^{-3} to 10^{-5} M); (2) an excess of nonreducible ions (e.g., 0.1 M); (3) a small amount of surface-active agent (e.g., gelatin); and (4) probably an excess of complexing agent. For accurate chemical analysis, it is necessary to select the most appropriate base solution for the mixture being considered. This problem is simplified by the half-wave potential tabulations included in most monographs on polarography and in the "Handbook of Analytical Chemistry."

The fundamental units of the apparatus used for polarography are shown in Figure 5.3 but as with all instrumental techniques, there are many different designs and a large number of modifications. In recording polarographs, the instrument covers the potential range in a matter of minutes and the current-voltage curves are automatically drawn on a chart recorder. In ac polarography, a small ac voltage is superimposed on the variable applied dc voltage to facilitate direct recording of a differential curve. In this way, the sensitivity of the technique is increased manyfold. Apparatus has also been designed in w which the whole selected potential range is swept during the lifetime of one mercury drop and the resulting current-voltage curves are displayed on a cathode ray oscillograph. In the square wave and pulse polarograph, a short square wave voltage pulse is imposed at regular intervals onto a linearly increasing dc voltage scan. By measuring the current during the last few milliseconds of the square wave, the recorded signal represents essentially only faradaic (i.e., diffusion) current.

With modified circuits the limits of detection can be as low as 10^{-7} to 10^{-8} molar (i.e., μg/liter range).

More than one determination may be achieved on a single polarogram but the preliminary treatment of samples can be complex and time consuming. Oxygen dissolved in electrolyte solutions is easily

reduced at the mercury cathode, and it is therefore necessary to remove oxygen by bubbling hydrogen or nitrogen through the solution for
10 to 15 minutes before running the polarograph, if determining the
metal content. At the same time, it should be noted that polarography
has been widely used to determine the dissolved oxygen content of
solutions.

To obtain suitable half-wave potentials for all the components of
a sample, two or more base solutions may be required. Further
complications arise if a major impurity has a lower half-wave potential than a minor impurity, for this reduces the sensitivity that can be
used to study the minor component. Finally, it is necessary to ensure
that the base solutions and temperatures used for the preparation of
calibration graphs and samples are as similar as possible.

The magnitude of the diffusion coefficients of species vary with the
temperature and environment, hence it is essential to correlate diffusion
current with concentration by means of an appropriate standardization
procedure. One widely used technique is to add several, known, small
amounts of the specified metal ion to a series of test solutions (i.e.,
base plus sample). Extrapolation of a plot of current versus standard
addition back to zero current allows calculation of the sample concentration.

With strict adherence to established technique, the reproducibility
of duplicate analyses may be as good as $\pm 2\%$. This precision compares
favorably with values quoted for spectrophotometric and atomic absorption methods (for the same concentration range) but as explained in
Chapter 2, accuracy and precision are two different terms, and many
factors contribute to the total error. Routine analyses are rarely performed under the ideal conditions required for optimization of precision,
and sampling or preparation errors can be far greater than the errors
associated with the analytic procedure. Accordingly, it may be more
realistic to assume that the percent relative error in most trace analyses
is of the order of $\pm 10\%$.

The technique known as <u>anodic stripping voltammetry</u> (ASV) is
proving to be of particular value in studies of metal ions present in the
parts per billion (μg/liter) region.

It may be considered as a modified form of polarography. The
instrumental unit contains an electrolysis cell made up of a working
electrode (Hg drop or Hg film on a polished carbon rod), an auxiliary
electrode (Pt), and a reference half-cell, connected to a power source
capable of applying a potential of 0 to 3 V. The electrodes are covered
by the aqueous sample, to which may have been added a base electrolyte

solution (to give electrical conductivity or change the chemical form of the metal ion species present). Dissolved oxygen is removed from the solution with a stream of inert gas (N_2 or He) and then a negative potential (sufficient to reduce the constituent(s) of interest, e.g., -1.0 V) is applied to the working electrode for a fixed period of time (the solution is stirred during this period).

At the end of the deposition cycle, the polarity of the electrodes is reversed and the potential is quickly returned to zero. Any deposited metals are oxidized in sequence of their half-wave potentials, and a plot of current versus voltage gives a series of peaks, the height of which is proportional to the amount in the initial sample.

Typical anodic stripping curves are shown in Figure 5.5.

The electrode systems currently available have some disadvantages which restrict their use in routine analysis. For example, a hanging mercury drop electrode (HMDF) provides relatively low sensitivity, and suffers from poor precision and resolution of neighboring waves. Thin mercury film electrodes (on Ni, Pt) offer improved sensitivity and

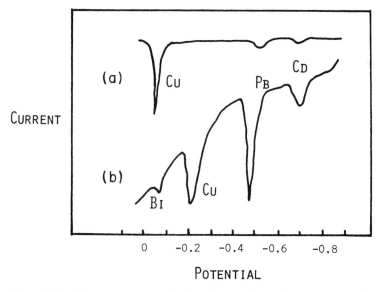

Fig. 5.5 Voltammograms obtained in studies of marine samples. Sweep rate 3 V/min. (a) Fish skeleton, citrate buffer, sensitivity 0.8 μA/cm; (b) seawater, pH 1.9, sensitivity 0.2 μA/cm. (From: Florence, T. M.: J. Electroanal. Chem. 35:237, 1972.)

resolution, but it is difficult to prepare films of uniform thickness and results tend to be less reproducible than that given by HMDE. Polished glassy carbon electrodes appear to have great potential, particularly if a mercury film is formed in situ. Mercury(II) ions are added to the sample and the potential used for deposition is set at a value which deposits both mercury and the trace metals. At the end of a run the mercury film can be removed by wiping with tissue paper. The technique is highly sensitive and gives excellent resolution of neighboring waves.

The range of elements amenable to study by this technique and the sensitivity of the procedure may be gauged from the seawater analyses recorded in Table 5.5.

Table 5.5

Analysis of Surface Seawater

| ELEMENT | CONCENTRATION[a] μg/liter | | | |
| | Soluble Fraction | | Insoluble Fraction[b] | |
	1[c]	2	1	2
Bi	0.21±0.02	0.043±0.02	0.011±0.002	<0.005
Sb(III)	<0.05	<0.05	–	–
Cu	9.8±0.7	0.93±0.10	0.29±0.01	0.11±0.01
Pb	1.04±0.05	0.42±0.04	0.21±0.03	<0.03
In	<0.05	<0.05	<0.005	<0.005
Cd	0.27±0.03	1.06±0.03	0.83±0.05	0.05±0.02
Tl	<0.2	<0.5	<0.05	<0.01
Zn	5.6±0.3	6.51±0.4	0.31±0.01	0.11±0.05

[a]Results are mean and difference of two analyses.

[b]Collected on a 0.45 μm Millipore filter. Based on data published by Florence, T. M.: J. Electroanal. Chem. 35:237, 1972.

[c]Sample numbers.

To work successfully in the parts per billion (μg/liter) concentration region one needs to be constantly suspicious of events or operations which may lead to contamination of the sample or losses of metal. Most of the problems associated with anodic stripping voltammetry arise predominantly from side effects such as adsorption on containing vessels or impurities in the distilled water and chemicals used to prepare standard solutions.

The technique is also limited at the moment in respect to the number of metals which can be evaluated, but where applicable it probably represents the most sensitive technique currently available. Sensitivity limits (in μg/liter) which have been proposed are of the order of 0.2 (Sn), 0.1 (Tl), 0.05 (Ag, Cd, Hg, Pb), 0.03 (Cu, In, Zn), 0.01 (Sb), and 0.005 (Bi). Comparison of these values with atomic absorption detection limits (Table 5.4) clearly illustrates the range of concentrations in which ASV will prove invaluable.

VI. DETERMINATION OF ION ACTIVITIES BY ION-SELECTIVE ELECTRODES

The use of glass electrodes and attendant meters to measure the pH of aqueous solutions is a well established technique. Of more recent origin is the introduction of electrodes which allow other ionic species to be examined with equal ease.

As shown in Table 5.6 electrodes are now available for determining a range of cationic and anionic species. To this list one can add solid metal electrodes (e.g., Ag probes for Ag^+ studies) and platinum electrodes for monitoring changes in redox potentials [e.g. changes in $Cr(VI)/Cr(III)$ ratios].

In brief, the measuring procedure is as follows: An ion-selective electrode and a suitable reference electrode (of fixed potential) are immersed in standard solutions of the ion of interest, and the millivolt potential developed between them is measured with a high-resistance potentiometer (pH meter). The readings obtained with the circuit shown diagrammatically in Figure 5.6 are plotted against concentration to give a calibration graph which can be used to interpret the readings obtained with test samples.

At this point, it should be emphasised that ion-selective electrodes measure the <u>activity</u>[*] of a given species.

---

[*]Activity, a = concentration, (mole/liter) x activity coefficient (f). An estimate of the magnitude of the coefficient (f) can be derived from equations such as $\log f = -Az^2$ ($I^{\frac{1}{2}}$) where the ionic strength (I) reflects the concentration (c) and charge (z) of all ionic species present ($I = \frac{1}{2} \sum_i c_i z_i^2$). A varies with temperature and dielectric constant of the medium.

Table 5.6

The Properties of Ion-Selective Electrodes[a]

Ion (X)	Range[b,c] pX[d]	pH	Interfering ions	Type
H^+	0-10	-	Na^+, K^+, Li^+	Glass membrane
Li^+	-	-	-	Glass membrane
Na^+	0-6	3-12	-	Solid state[e]
K^+	0-6	-	Rb^+, Cs^+	Liq. membrane[e,f]
Ag^+	0-7	0-14	Hg^{2+}	Solid state[e]
Ca^{2+}	0-5	5.5-11	Divalent cations	Liq. ion exchange
Cu^{2+}	0-8	0-14	Ag^+, Hg^{2+}, Fe^{3+}	Solid state
Cd^{2+}	1-7	1-14	Ag^+, Hg^{2+}, Cu^{2+}	Solid state
Pb^{2+}	1-7	2-14	Cd^{2+}, Fe^{3+}, Ag^+, Hg^{2+}, Cu^{2+}	Solid state
F^-	0-6	0-8	OH^-	Solid state
Cl^-	0-4	0-14	S^{2-}, CN^-, I^-, Br^-, $S_2O_3^{2-}$	Solid state
Br^-	0-5	0-14	-	Solid state
I^-	0-7	0-14	Ag^+, S^{2-}	Solid state
CN^-	2-6	0-14	Ag^+, S^{2-}, I^-	Solid state
CNS^-	0-5	0-14	I^-, Cl^-, Br^-, $S_2O_3^{2-}$	Solid state
NO_3^-	1-5	2-12	ClO_4^-, I^-, ClO_3^-, Br^-, S^{2-}	Liq. ion exchange
ClO_3^-	1-5	4-10	OH^-	Liq. ion exchange
BF_4^-	1-6	2-12	I^-	Liq. ion exchange
S^{2-}	0-7	0-14	Ag^+	Solid state

[a] Based mainly on data published by Orion Research Incorporated.

[b] Some electrodes will detect activities lower than the pX values noted, but the total concentration must be in the range indicated.

[c] pX values > 5 represent sub-ppm concentrations.

[d] pX = -log (molar concentration).

[e] Glass membrane types also available; H^+ interferes hence pH > 5 before use.

[f] Philips product.

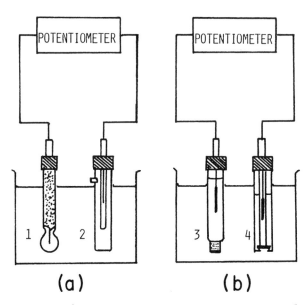

Fig. 5.6 Circuits for potentiometric measurement. (a) Glass membrane electrode (1); calomel reference electrode (2). (b) Ion-selective
electrode (liquid membrane) (4); silver, silver chloride reference
electrode (3).

Electrode descriptions: (1) Bulb of pH-sensitive glass contains an
Ag, AgCl electrode in 0.1 M HCl; (2) Paste of Hg, HgCl$_2$, KCl in contact with saturated KCl solution, salt bridge (e.g., fiber) contact with
test solution; (3) Ag wire coated with AgCl immersed in saturated KCl,
solution contact via inbuilt salt bridge; (4) Ag/AgCl electrode immersed
in liquid exchanger of appropriate (e.g., Ca) ionic form, a membrane
serves as interface between test solution and exchanger medium

The relationship between ionic activity and the potential developed
by an ion-selective electrode is logarithmic.

$$E = E_a + 2.3\frac{RT}{nF} \log a$$

where a is the activity of the ion in solution, E is the measured total
potential of the system, E_a is the portion of the total potential due to
choice of reference electrode and internal solutions, and 2.3RT/nF is
the Nernst factor [59.15/n mV at 25°C, 74/n mV at 100°C] in which R
and F are constants, T is temperature (K), and n is the charge of the
ion. From the above equation, a tenfold change in ionic activity causes
a change in electrode potential (at 25°C) of 59.16/n mV.

For a given concentration of the species of interest, the response of the electrode can thus vary with the ionic strength and temperature of the solution. In the presence of a complexing agent the response is further altered due to coordination. For example, the potential of a copper electrode is determined by the activity of the hydrated copper(II) ions in solution; in the presence of a ligand (e.g., NH_3) this value is itself determined by the stability of the complex and the concentration of ligand, i.e., $[Cu^{2+}] = \beta_x[CuL_x]/[L]^x$.

All electrodes produced to date respond to ions other than the one they are intended to measure (the primary ion). Details of potential interferants are usually supplied with the electrodes and means of mathematically correcting for some effects may be included. However, to produce meaningful measurements one must know not only the nature of all the ions in the solution but also their approximate concentrations.

The general properties of a number of commercial electrodes are summarized in Table 5.6, and it will be noted that three distinct types of electrodes are identified:

1. Glass membranes. The sensor unit is a thin bulb of glass and selectivity depends on glass composition. Most widely used in H^+ studies; other units are useful for Li^+, Na^+, K^+, or Ag^+ determinations.

2. Solid state membranes. Sensor unit is a solid disk containing an ionic conductor. Widely used are silver halides or sulfide, since membranes containing these materials develop a potential controlled by the silver ion activity in the test solution. The latter can be determined by solubility considerations, e.g., in the presence of halide ions $a_{Ag} = K_{AgX}/a_X$, where K_{AgX} is the appropriate solubility product; and response thus varies with halide activity. Sensors containing a mixture of silver and other metal sulfides (e.g., Cu^{2+}, Cd^{2+}, Pb^{2+}) respond to changes in the metal ion concentration, since the cation controls the S^{2-} activity and thus, indirectly, a_{Ag}. The lanthanum fluoride single crystal electrode (for F^-) is most selective of all electrodes and is widely used in pollution monitoring.

3. Liquid membranes. In these electrodes, a liquid ion exchanger is converted to the ionic form of interest and a layer of this organic liquid is separated from the test solution by a sintered glass disk or plastic membrane (e.g., a calcium electrode may have a calcium organophosphorus compound as its inner filling solution). A wide range of electrodes have been produced based on this principle.

For regular monitoring of the activity of species, direct measurement by means of ion-selective electrodes is an attractive proposition. To obtain valid results, however, it is necessary to minimize interference effects and ensure that the calibration standards are similar in composition to the test samples.

Ionic strength effects can be swamped out by adding enough inert electrolyte to the sample (and standards) to produce a large increase in the total electrolyte concentration. Through careful pH adjustment one can control protonation of ions of interest (e.g., S^{2-}, HS^-, H_2S) and minimize some complexation reactions. Another mode of correcting for complex formation involves adding a species which has a greater affinity for the interfering ion (e.g., fluoride present as an aluminium complex can be released by adding citrate ions).

It should be noted that in order to achieve more than $\pm 10\%$ precision with respect to concentration values, potential readings should be consistent to better than ± 2 mV.

For measuring total concentrations, (as distinct from activities) one can often obtain far more accurate results by using the electrode system solely to detect the end-point in a titration of the species of interest with some appropriate reagent (e.g., Ca^{2+} with EDTA).

The range of applications is now quite extensive and the literature contains many methods for evaluating components regarded as pollutants in water studies.

FURTHER READING

Water Analysis

Most libraries possess a range of books on water pollution, water analysis, and specialized techniques which can be used to expand all aspects covered in this chapter. For example:

American Public Health Association, "Standard Methods for the Examination of Water and Wastewater," 14th ed., Am. Public Health Assoc., Washington, 1976.

American Society for Testing and Materials, 1970, Annual Book of ASTM Standards, Part 23: Philadelphia, Am. Soc. of Testing Materials.

Brown, E., Skougstad, M. W., and Fishman, M. J.: "Methods for Collection and Analysis of Water Samples for Dissolved Minerals and Gases: Techniques of Water-Resources Investigations of the U.S. Geol. Survey," Book 5, Chapter A1, Washington, D.C., Superintendent of Documents, U.S. Govt. Printing Office, 1970.

Ciaccio, L. L., ed.: "Water and Water Pollution Handbook," Vols. 1-4, Dekker, N.Y., 1971-3.

Environmental Protection Agency, "Methods for Chemical Analysis of Water and Wastes," E.P.A. Water Quality Office, Cincinnati, 1971.

Golterman, H. L., ed.: "Methods for Chemical Analysis of Fresh Waters," Intern. Biological Programme, Handbook No. 8, Blackwell Scientific Publications, Oxford, 1971.

Specialized Monographs

Boltz, D. E., ed.: "Colorimetric Determination of Non-metals," Wiley, N.Y., 1958.

Dean, J. A., and Rains, T. C.: "Flame Emission and Atomic Absorption Methods," Vols. 1-3, Dekker, N.Y., 1969.

Durst, R. A., ed.: "Ion Selective Electrodes," National Bureau of Standards Special Publication, Washington, 1969.

Heyrovsky, J., and Kuta, J.: "Principles of Polarography," Academic Press, N.Y., 1966.

Ramirez-Munoz, J.: "Atomic Absorption Spectroscopy," Elsevier, Amsterdam, 1968.

Sandell, E. B.: "Colorimetric Determination of Traces of Metals," 3rd ed., Wiley, N.Y., 1959.

Reviews and Articles

Annual Reviews, Anal. Chem., April issues, Vol. 43, 1971; Vol. 45, 1973; Vol. 47, 1975.

PRINCIPLES OF EMISSION SPECTROSCOPY,
GAS CHROMATOGRAPHY, MASS SPECTROMETRY,
AND NEUTRON ACTIVATION ANALYSIS

I. EMISSION SPECTROSCOPY

A. Radiation Emission

As indicated in Chapter 4 (Section IV), chemical systems can be excited through the absorption of energy, and when they revert to the original or ground state, the additional energy has to be lost either as heat or in the form of radiation.

The multiple energy transitions which a system can undergo are characteristic of the species being excited; hence any emitted radiation is also characteristic of the system.

By means of a monochromator, the sample emissions can be dispersed into components which differ in wavelength. Measurement of the wavelengths emitted permits identification of the excited material, while measurement of the intensity of selected wavelengths can be used to estimate the amount of a particular species.

Several modes of excitation of the sample are available and this leads to subclassifications such as x-ray emission spectroscopy (discussed in Section IV.B of Chapter 4) flame photometry, fluorimetry (Section IV of Chapter 4) and emission spectroscopy.

As an analytic technique, emission spectroscopy is mainly limited to individual elements, e.g., all the metals and certain of the nonmetals such as carbon, boron, and phosphorus. There is little opportunity to analyze for molecular species, since at the high temperature of excitation required, chemical bonds are too easily broken.

As illustrated in Figure 6.1, the basic equipment required for this technique consists of three interrelated units: (1) a means of exciting

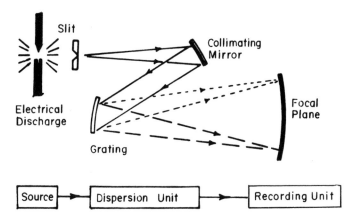

Fig. 6.1 Basic system for emission spectroscopy. (Reproduced from
Pickering, W. F.: "Modern Analytical Chemistry," Dekker, 1971 with
permission of the publisher.)

the sample, (2) a means of dispersing the emitted radiation into constituent frequencies, and (3) a means of detecting and recording the dispersed radiation.

The optical dispersion systems produce sharp images of the initial
slit at the focal plane of the equipment, and on a photographic plate these
images appear as a series of parallel lines, each line corresponding to
a particular energy transition in an excited atom. The characteristic
wavelengths of emitted radiation are accordingly often referred to as
lines.

The number of lines observed depends partly on the magnitude of
the initial energy input and partly on the number of transitions which are
statistically feasible.

The intensity of a particular emission is determined by the number
of atoms in the appropriate excited state (N_j). As noted previously
(Chapter 4) only a small proportion of the atoms present become excited
on exposure to, say, heat energy

$$\frac{N_j}{N_o} = \left[\frac{P_j}{P_o} \exp \frac{-E_j}{kT} \right] <<< \text{ unity (e.g., } 10^{-2} \text{ to } 10^{-20})$$

For a particular transition, the probability terms (P_j and P_o) and
energy term (E_j) are fixed, hence the intensity becomes proportional

to the total number of atoms (i.e., concentration of element in sample) and the temperature. Accordingly, in quantitative studies which rely on intensity comparisons, it is essential to fix the temperature or provide means of correcting for temperature variations.

Since F_j for the most probable (hence most intense) transition is different for each element, at a given temperature, similar total concentrations of different elements yield vastly different emission intensities. Conversely, to obtain a detectable signal, different total concentrations are required (for example, in a flame photometer, the excitation of 0.001 ppm of sodium may yield a similar meter deflection to 10 ppm of lead).

During excitation, some electrons can be totally ejected (creating ions). The degree of ionization of any element increases with temperature, and a point can be reached where this process so depletes the number of available neutral atoms that there is a weakening of emission intensity.

Another factor which influences the observed intensity of radiations is the phenomena known as self-absorption. At any time the number of atoms in the ground state (N_0) greatly exceeds the number of excited states, and as the total concentration increases, the probability of emitted radiation meeting lower energy atoms before entering the dispersion system is markedly increased. Ground state atoms absorb characteristic emission; (compare atomic absorption) and this self-absorption leads to a decrease in the emitted intensity.

A final factor influencing the intensity of an emission is selective volatilization. In a mixture of elements, some may be almostly completely vaporized before others begin to distil, leading to atom concentrations which differ with time of excitation. The vapors excited are thus not truly representative of the composition of the base sample.

Despite these limitations, emission spectroscopy offers extreme specificity and sensitivity in identifying and measuring concentrations of elements. All ranges of concentration can be handled in qualitative work, down to less than one part per million in many cases, and analyses may be performed on samples as small as 1 mg. Another advantage, a result of the specificity of the method, is the possibility of doing a nearly complete elemental analysis on a single sample of complex material.

B. Flame Photometry

A hot flame provides sufficient energy to excite many elements. Any fuel gas with a high combustion temperature can be used, but the most common fuels used are propane, acetylene, and hydrogen. When burned with air, these gases yield flames giving temperatures between 1700° and $3200^{\circ}C$. For many elements, the atom formation process is enhanced by substituting nitrous oxide as the support gas (Section IV.A of Chapter 4).

For each element being studied, it is desirable to ascertain the best flame conditions by varying the components of the flame and by varying the pressures of the gases being supplied to the burner. Once optimum conditions have been determined, gas pressures must be maintained constant during all tests and during preparation of calibration curves in order to maintain uniform excitation of the samples. Since the flame has different temperatures and radical concentrations in different zones, the same area of the flame should be focused on the entrance slit for any series of determinations.

A major advantage of flame excitation is that the sample can be introduced to the burner in the form of an aerosol, a steady flow of solution being achieved by incorporating a nebulizer in the burner unit. The type of assembly shown in Figure 6.2 feeds only the finer droplets, i.e., a homogeneous fog, to the burner; another type of unit injects the entire spray directly into the flame.

Because of the comparatively low excitation energies being applied, the flame emission spectrum is often reasonably simple and filters can isolate satisfactorily the characteristic radiations of a particular element. In more sophisticated equipment, monochromators based on a prism or diffraction grating are used.

The intensity of the selected radiation is usually measured directly by means of a phototube or photocell coupled with an amplifier and either a meter or recorder.

The majority of errors in flame photometry arise from phenomena that develop in the flame.

Two common causes of nonlinear calibration curves (i.e., intensity of radiation plotted against solution concentration) are self-absorption and ionization.

Chemical interference arises from the formation in the flame of condensed phases which are difficult to volatilize and dissociate into free metal atoms. For example, a number of elements form oxides of the type MeO, and the emissions from these species (if any) are in the form of broad bands rather than sharp lines. (The use of nitrous oxide-acetylene flames has eliminated many of these difficulties.) Salts introduced with the sample can be responsible for the formation

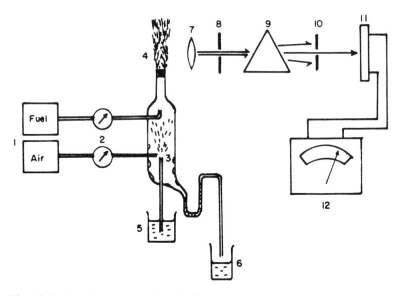

Fig. 6.2   Basic components of a flame spectrophotometry unit: 1.
Cylinders of fuel and air (or O_2); 2. pressure regulating valves and
flowmeters; 3. spray chamber; 4. burner; 5. test solution; 6. spray
chamber drainage system; 7. focusing lens; 8. entrance slit; 9.
wavelength selector; 10. exit slit; 11. photoelectric detector; and
12. recording meter. (Reproduced from Pickering, W. F.: "Modern
Analytical Chemistry," Dekker, 1971 with permission of the publisher.)

of refractory species, for example, the presence of sulfate, phosphate,
and several other oxyanions causes a decrease in emission intensity
from solutions containing the alkaline earth elements. In many determinations this problem can be minimized by using releasing agents
(i.e., addition of a competing ion which combines preferentially with
the interferent species) or by protective chelation (i.e., formation of
a stable complex of the element of interest).

     The composition of the test solution can influence the efficiency of
atom formation significantly. In addition, it can influence the fraction
of sample admitted to the flame as a fog.

     To minimize chemical interference effects and gain greater sensi-
tivity, it is often recommended that the species of interest be isolated
by solvent extraction before being introduced into the flame. The use
of an organic solvent adds another variable to the system.

Flame photometry has proved to be a very sensitive technique for the determination of metals such as the alkali metals which are easily excited, but less satisfactory for metals which require high-energy inputs. With better class equipment the detection limits can be as low as 1 μg/liter and regularly elements are determined at the μg/ml level. However, it is essential to ensure that the composition of the standards used in calibration is nearly identical in nature to that of the unknown.

C. Electrical Excitation

Electrical discharges are very effective in volatilizing and exciting solid samples. Depending on their specific properties and means of generation, they are classified as arcs, sparks, or simply discharges.

An arc is an electrical discharge which has to be initiated by an auxiliary spark or by momentary mechanical connection across the electrode gap. Very minute amounts of material are volatilized and excited, and this mode of sample treatment has proved extremely valuable for qualitative analysis. However the electrical circuit has to be well stabilized to be satisfactory for quantitative analysis. The arc discharge tends to wander across the sample, with accompanying changes in current flow and temperature; this leads to variations in emission intensity with time and nonuniform vaporization of the sample. (Arcs operate at temperatures between 3000° and $8000^{\circ}K$, the temperature depending on the applied voltage and the nature of the vapors in the gap). Undesirable effects can be minimized by technique variations such as gas sheathing and low-pressure stabilization but for quantitative studies, spark excitation is normally preferred.

Spark discharges jump electrode gaps unassisted and give spectra rich in high-energy lines. To initiate a spark discharge a voltage of 10 to 30 kV is applied across two electrodes, one of which is the sample. Provision is made to restrict the time of each discharge, so that excitation consists of a series of sparking operations. Each current pulse takes material from the electrodes into the spark gap where the atoms are excited by collision with high-energy electrons.

Another form of high-temperature excitation (e.g., $> 8000^{\circ}K$) which is receiving much attention is the plasma torch. In the dc-arc type of plasma jet, a closed chamber contains an anode at one end and at the other end a cathode with a small hole in it. A stream of ionizable gas (argon) is admitted tangentially through the chamber wall. When an arc is struck, there is a thermal pinch effect and hot plasma is ejected through the opening. The plasma has the general appearance of a bright flame and systems to be excited are usually introduced into the plasma in the form of an aeosol.

The radiations emitted in the electrical discharges are focused on
a fine slit and the transmitted radiations are dispersed either by means
of a prism or a diffraction grating. The diffracted radiations are
brought to focus at a focal plane for recording (Fig. 6.1).

The detector-recorder system used for almost all qualitative, and
much quantitative work, is a photographic emulsion. The exposure
time has to be carefully adjusted and developing procedures must be
standardized. To permit identification of the frequency of the constituent radiations, spectra of known pure materials such as copper or
iron are photographed on the same plate. For quantitative determinations, the density of the photographic image of selected lines is measured by means of a microphotometer. The measured density is then
related to concentration by means of calibration graphs prepared from
standard samples of the same material.

For routine quantitative analysis, photoelectric measurement has
been widely adopted. Photomultiplier tube detectors are positioned
behind slits cut in an opaque barrier located along the focal plane, each
slit being positioned to correspond to some characteristic line of different elements. The output of a detector is integrated over a period
of 25 to 40 seconds, and the intensity of the line of interest is measured
by comparing the voltage accumulated across a capacitor with the voltage
developed by a selected reference line.

Almost all quantitative determinations are based on the comparison
of the intensities of specially selected sets of lines known as homologous
pairs. These are lines of about the same wavelength whose sensitivity
to the applied excitation energy is of the same order of magnitude, so
that variations in procedure or conditions produce the same effect on
both lines.

The electrodes used in emission spectroscopy are generally made
of graphite, although metal samples may be cast into disks or rods and
used directly. For studies on powdered materials, several techniques
are available. The powder may be placed in a hollow in a graphite
electrode or pressed into a pellet suitable for direct arcing and sparking.
However, since the nature of the matrix can influence the intensity of
the emitted lines, it is essential that the general composition and physical properties of standards and unknowns should be the same or closely
similar.

The range of applications is extremely broad, the technique having
been used successfully for the analysis of rocks, minerals, metal
alloys, and commercial products of all types. Multiple analyses can
be performed on milligram amounts of sample and many of the elements

determined readily by this technique are difficult to analyze by alternative procedures. The greatest advantages, however, are the ease, speed, and accuracy with which qualitative analyses of a wide range of materials can be performed.

II. GAS CHROMATOGRAPHY

A. Factors Influencing Separations

The determination of organic compounds by the technique known as gas chromatography has been outlined in general terms in a preceding chapter (Section III.A of Chapter 3), with the components of the equipment being shown in Figure 3.7. In this chapter, the aim is to consider some of the basic principles and difficulties associated with the technique.

Components of a mixture are separated by differential migration through a packed column, and an examination of a block diagram of the analytic system (Fig. 6.3) shows that the experimental variables include the velocity of carrier gas, temperature, nature of the material in the separation column, and type of detection system. These variables have to be manipulated to achieve selective distribution of sample components between the mobile gas phase and the column material, with subsequent identification of the separated species.

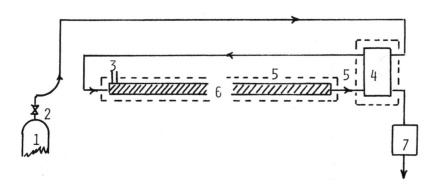

Fig. 6.3 Basic components of a gas chromatograph: 1. Cylinder of carrier gas; 2. gas regulating valve; 3. sample introduction system; 4. differential detector; 5. thermostats; 6. chromatography column; 7. gas flowmeter. (Reproduced from Pickering, W. F.: "Fundamental Principles of Chemical Analysis," Elsevier, 1966, with permission of the publisher.)

When a sample is admitted to the top of a packed column, an equilibrium situation develops. Some of the sample vapor is distributed into the solid phase, the rest remains in the gas phase. As the mobile phase moves into the next small segment of the column, it carries with it the vapor from the first hypothetical segment. A new distribution occurs and more vapor is transferred to the column packing. The moving gas takes the residual vapor along to the next zone of fresh packing, and the process continues along the length of the column. In the upper segment of the column, the reverse distribution occurs as fresh gas comes in contact with the initial sample zone. The new vapor phase, as it moves along, comes to equilibrium with the various fixed components in the various segments.

For each component of the mixture, within each hypothetical segment, the equilibrium distribution may be of the type $C_1 = k(C_2)^n$ where C_1 is the concentration in phase 1, C_2 is concentration in phase 2, and the power n can have integral or fractional values.

If conditions are arranged to ensure that the proportionality constants (k) for individual components of a mixture have different numerical values, then each component moves through the column at a different rate and emerges at a different time. Depending on the nature of the distribution isotherm, the detector response for each emerging peak tends to resemble the shapes shown in Figure 6.4.

For effective separation it is desirable to arrange conditions so that symmetrical peaks are observed.

In gas chromatography this ideal can be achieved by coating the column material with a substance which acts as a solvent for the sample components, because the dissolution of a vapor in a liquid normally yields a distribution isotherm of the type $C_1/C_2 = k$. This desired linear relationship is less likely to be achieved if the sample species are held to the solid by adsorption forces.

The sharpness of an elution peak depends on the number of equilibrium processes that occur during the passage of the solute through the column, and on the time required to elute the material from the column.

The efficiency of a column in effecting separation depends on the separation achieved in each individual equilibration process (separation factor), and the total number of equilibration processes (Fig. 6.5).

The separation factor (α) can be defined as $\alpha = k_A/k_B$ where k_A represents the distribution coefficient of substance A between the two phases and k_B represents the corresponding distribution coefficient for substance B. $[k = C_L V_L/C_G V_G = K V_L/V_G$ where C_L and C_G are

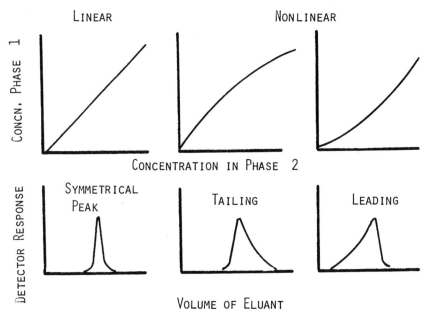

Fig. 6.4 Diagram showing relationship between type of distribution isotherm and shape of elution curve. (Reproduced from Pickering, W. F.: "Fundamental Principles of Chemical Analysis," Elsevier, 1966, with permission of the publisher.)

the concentrations in the liquid and gas phase, V_L and V_G are the respective phase volumes. The magnitude of $K(=C_G/C_L)$ is determined by the solute, solvent and temperature].

On gas chromatograms, the separation factor, for two components, is equal to the ratio of the two retention volumes (i.e., volume of gas required to elute the components from the column). The distribution coefficients are also a measure of relative volatility, hence the separation factor can be defined as $\alpha = V_{g1}/V_{g2} = a_2 p_2^0/a_1 p_1^0$ where p_1^0 and p_2^0 represent the vapor pressures of solutes 1 and 2, respectively, at a given temperature, and a_1, a_2 represent primarily the activity coefficients of the sample components in the liquid phase. (If species 2 is an arbitrarily selected standard compound, V_{g1}/V_{g2} is also termed as a relative retention value.)

When activities equal unity (i.e., ideal dilute solutions) the relative retention value is determined solely by the solute vapor pressures. Conversely, when dealing with materials having similar vapor pressures

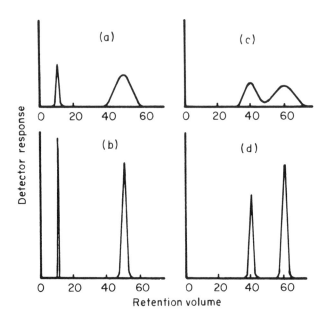

Fig. 6.5 Diagram illustration the effect of separation factor (α) and number of theoretical plates (n) on the degree of separation of two substances having linear distribution isotherms: (a) $\alpha = 5$, $n = 100$; (b) $\alpha = 5$, $n = 1000$; (c) $\alpha = 1.5$, $n = 100$; (d) $\alpha = 1.5$, $n = 1000$. (Reproduced from Pickering, W. F.: "Modern Analytical Chemistry," Dekker, 1971 with permission of the publisher.)

at a given temperature (i.e., similar boiling points), separation depends on alteration of the a_2/a_1 ratio by solute-solvent interaction.

In general, coatings of nonpolar liquid phases are nonselective (i.e., solute volatility is determined mainly by the vapor pressure) and elution from a column occurs in the order of the boiling points of the solutes. With polar liquid phases, polar components of the mixture are preferentially retarded, and the polarity of the liquid phase is important. Interaction is also a function of chemical type, e.g., aromatic compounds are selectively detained on liquid phases containing aromatic rings.

As shown in Figure 6.5, the detection of eluted components is improved by having values of α much greater than unity, hence care is needed in selecting the solvent phase and the temperature of operation. It may also be noted from this diagram that sharpness varies with n, and for comparison purposes, it is accepted practice to calculate the number of distributions (n) from the sharpness of the peak, e.g., n = $(4R_V/\Delta_V)^2$ where R_V is retention volume* for the peak and Δ_V is a measure of the width of the peak base.

The number of equilibrations in a column of given dimensions depends in a rather complicated way on a large number of factors, including uniformity of column packing, flow rates, and kinetic effects associated with the interphase distribution.

The parameters which contribute to the length of column required to establish equilibrium between two phases (known as the height equivalent to a theoretical plate, HETP) have been divided by Van Deempter into three groups: eddy diffusion, molecular diffusion, and resistance to mass transfer. The magnitude of the last two of these terms depends on the velocity of the mobile phase (u):

$$HETP = A + \frac{B}{u} + Cu$$

The term A varies with average particle size of the packing and uniformity of packing.

Term B reflects the tortuosity of the interparticle spaces and the diffusion coefficient of the species in the gas phase.

Term C varies as the square of liquid film thickness, inversely as the diffusion rate in the liquid phase, and nonlinearly with the distribution coefficient.

$$\text{The number of equilibrations, } n = \frac{\text{length of column}}{\text{HETP}}$$

Detailed mathematical analyses of chromatographic processes have led to the postulation of more rigorous equations, but the simplified Van Deempter equation is sufficient to indicate that due to the counteracting effects of the B/u and Cu terms, there exists an optimum flow rate of mobile phase. It is also apparent that columns will vary in performance through variations in packing efficiency and coating thickness.

For the separation of a given group of solutes, the operator can vary the nature of the fixed phase, the particle size and shape of the solid support, the nature and velocity of the mobile phase, and the temperature. This provides great versatility, but the elucidation of optimum conditions for a given application can involve much trial and error experimentation.

---

* Retention volume = retention time x flow rate of gas.

B. Gas-Solid Chromatography

For the separation of gases and vapors of low boiling point (e.g., CO, CO_2, O_2, NO, hydrocarbons to C_5), uncoated solid adsorbents may be used in the instrument column.

The alternative packings available include activated carbon, alumina, silica gel, and molecular sieves. The columns are commonly about one-fourth-in. in diameter and can vary in length from 2 to 20 ft or more. The tube may be made of glass, copper, aluminum, or stainless steel, and is often formed into a coil shape to facilitate mounting in the temperature control oven.

Small volumes of gaseous sample are introduced at the top of the column by means of a syringe or sample loop and are eluted through by a stream of carrier gas. The nature of the gas used (e.g., nitrogen, argon, carbon dioxide, hydrogen, or helium) depends partly on the composition of the sample and partly on the type of detector used to monitor the eluted components.

Even with selection of optimum gas flow rates and temperature of operation, resolution can be unsatisfactory because gas-solid chromatography usually leads to asymmetric elution peaks, and the relatively long tails can cause significant peak overlap.

Qualitative identification of the eluted species is usually based on retention volumes, through comparison with the behaviour of standard gases when using the same operating conditions. Quantitative measurements are based on measuring the area under a given peak. The relationship between area and concentration is established by inserting known amounts of pure gas or vapor into the instrument.

Gas-solid chromatography is somewhat limited by the small range of adsorbents available, the nonlinear nature of the distribution isotherms, and the need to replace or degas the adsorption column repeatedly.

C. Gas-Liquid Chromatography

The only variation in equipment required for gas-liquid chromatography is a change in the nature of the column packing. A solid support composed of particles of uniform size (e.g., 80 to 100 BSS mesh) is coated with a thin film of a liquid phase. The film must be even (to prevent gas adsorption on the solid and tailing effects) and as thin as feasible. The coatings must be completely nonvolatile (at the temperature of operation) and are normally chemically inert. The amount of coating material used is generally between 5 to 25% by weight.

Widely used support materials are diatomaceous earth and crushed firebrick, while for particular applications other materials, such as glass powder, sodium chloride crystals, metal helices, or granular polytetrafluorethylene have been recommended.

Broadly speaking, the success of the column as a separation medium depends on the choice of coating material. Besides possessing an extremely low vapor pressure ($<10^{-4}$ torr) at the operating temperature, the coating has to be thermally stable, be reasonably fluid, and possess some solvent powder for the mixture to be examined.

The large number of liquids suitable for use as stationary phases provides the chemist with a bewildering range of possibilities, but since many possess similar chemical and physical properties a significant proportion of columns are coated with one of the following materials: Dimethylformamide (20°), polyglycols (100°), benzyldiphenyl (120°), dinonylphthalate (130°), squalene (150°), diglycerol (150°), polyethylene glycol adipate (150°), apiezon grease M (150°), or L (230°), carbowax (250°), or silicone oils and greases (200 to 400°). The numbers in parentheses indicate the operating temperature above which loss of the stationary phase as vapor becomes excessive.

The choice of coating is determined by the chemical nature of the compounds to be separated (i.e., variation in solubility or activities) and the temperature required for fast, efficient separations.

The most suitable operating temperature for a gas-liquid chromatography unit is that which yields sufficient vapor pressure of the solutes to ensure rapid transit through the column, while retaining the ratio of solute vapor pressures (i.e., p_2^o/p_1^o) at a value which permits separation of the components.

With samples having components differing greatly in boiling point, it has been found desirable to temperature program. This procedure involves separating the more volatile components at a temperature which gives a reasonable elution time (e.g., several minutes).

The column temperature is then increased and the next group of components eluted. This procedure is repeated at higher temperatures to accelerate movement of the highest boiling point components. The temperature program may be arranged to give either a linear increase with time or a stepwise operation. The best program is that which separates all components as sharp peaks in a minimum of time.

Temperature control is equally important in two other parts of the gas chromatography unit.

To minimize peak broadening, the sample mixture should reach the top of the column as a compact plug of vapor. The injection port into which the mixture is admitted (e.g., few μl from a microsyringe) has, therefore, to be at a relatively high temperature to ensure rapid and complete vaporization. The temperature selected should be consistent with the thermal stability of the sample, although a modern trend with materials of very low volatility is to cause pyrolysis in the injector zone, and subsequently examine the fragments produced.

The other zone requiring temperature control is the detector unit, since it is necessary to prevent condensation of separated constituents as they pass through the detector. The sensitivity of katharometers (i.e., thermal conductivity detectors) decreases with temperature, hence the optimum temperature for this type of unit lies just above the boiling point of the highest boiling sample component. With other detectors (e.g., flame ionization types), maintenance of the vapor state in the feed lines is the prime consideration.

The selection of the detector system depends on the nature of the species being separated and the sensitivity required. A range of detectors is now available. Some have general applicability (e.g., thermal conductivity, flame ionization); others are more selective in their action (e.g., electron capture detectors for chlorinated compounds or alkali flame ionization for phosphorus compounds). Use of an inappropriate detector can result in nonobservance of peaks or poor precision in peak area measurements.

Gas-liquid chromatography is an extremely popular analytic tool because of its speed, simplicity, and versatility. However, it does possess a number of limitations. It is restricted to separations of volatile substances, there is a tendency for the stationary phase to bleed out of the system as temperatures are raised, and selection of column packings is somewhat empirical. More important is the fact that detection systems are usually nonselective. Similarity of retention behavior with that of a known sample, using several different columns at more than one temperature, is almost certain proof of identity. Unfortunately, this method is time consuming and requires a very extensive stock of pure chemicals. Absolute proof of identity requires secondary examination of the isolated species, e.g., by infrared spectroscopy or mass spectrometry.

Another alternative is to calculate relative retention volumes of the separated materials (using some internal standard) and compare them with literature values.

For quantitative studies the principle aim is to relate the area under an elution peak with the concentration of component through preliminary calibration studies, using varying volumes of pure compound. The data can lack precision due to the difficulty in accurately introducing small (i.e., μl) volumes of standards (or samples) and in accurately measuring areas. The calibration curve is not necessarily linear and the slope varies from compound to compound. Despite these limitations the method is usually more accurate than alternative procedures.

Clearly, gas chromatography can be used only for gases and volatile substances, and while the range is being extended continually through the use of high column temperatures, or formation of sample derivatives of greater volatility, or classification of pyrolysis patterns, there still remains a large number of compounds which cannot be studied by this technique. These are preferably separated by other forms of chromatography, and the approach of greatest potential for regular routine studies is probably high-pressure liquid chromatography.

III. HIGH-PRESSURE LIQUID CHROMATOGRAPHY

With the development of commercially available high-pressure liquid chromatography (HPLC) systems, chemists can now examine mixtures of nonvolatile species with similar facility to gas chromatography. Basically, the apparatus consists of a column, a hydraulic system, a detector, and a chart recorder to plot the elution pattern. With this equipment, molecular species can be separated at great speed and with high sensitivity, provided that they are soluble in a liquid phase and a suitable column packing is available.

The main distinguishing feature of HPLC is the use of pressure to pump the liquid mobile phase through the separation column. The columns used are generally small in bore and are packed with very fine particles (to yield a high surface area of contact). To produce a constant flow through extremely fine material, inlet pressures of up to 8000 psi are sometimes required. With other columns, efficient operation may be obtained using much lower pressures (e.g., 200 psi).

The column packing has to be capable of withstanding the pressure applied, but in nature, it can be an adsorbent, a liquid coated on a solid, an ion-exchange material, or a polymeric gel.

To get the best results it is important to choose the most appropriate type of packing. For example, if separation is to be based on molecular size (hence molecular weight) the packing should be a polymeric gel. For studies of ionic species in aqueous solutions, ion-exchange materials yield the best results. For nonionic species, the column packing is an adsorbent or a coated material. Packings available include etched glass beads, porous silica, pellicular ion exchange, silica, alumina, polyamide (coating on glass bead), and cross-linked polystyrene. Normally the stationary phase is polar and the mobile phase nonpolar in nature.

The solvent used must be degassed since air bubbles tend to block the column. The choice of solvent is determined in part by its ability to dissolve sample components, in part by its effect on the detector unit. To facilitate some separations, it is desirable to vary the polarity of the mobile phase during development through the addition of another solvent in controlled amounts.

The mixtures to be studied (μl quantities) are admitted to the top of the column either through a high-pressure inlet port or by temporarily stopping or diverting the liquid flow. On elution, the concentration of the separated components can be of the order of nanograms per milliliter, hence the detector systems need to be sensitive to such low levels, be fast in response, and have a small dead volume. The most widely used detector monitors changes in the absorption of a particular wavelength of ultraviolet light. Other detectors are based on changes in refractive index, electrical conductivity, heat of adsorption, or intensity of fluorescence. Modifications of flame ionization detectors or polarographic cells are being applied in research models.

The peaks appearing on the resultant chromatograms are identified by using an internal standard, or by comparing the retention volumes of the unknown peaks and standard compounds, or by collection of fractions followed by identification by infrared absorption or chemical methods (note similarity to gas chromatography).

With appropriate choice of conditions separation efficiency is generally high (e.g., 5000 theoretical plates per meter of column). The parameters contributing to the height of theoretical plate include those considered in the Van Deempter equation (Section A) plus corrections for coupling, turbulence, etc. Accordingly, a basic equation may be proposed in which the total HETP includes terms related to the speed of mass transfer in the mobile and stationary phases (respectively), and terms reflecting molecular diffusion caused by variations in fluid velocity occurring in the longitudinal direction and in the intergranular pores.

The technique has already been adapted to a number of pollution studies and the number of applications can be expected to increase significantly in the future.

IV. MASS SPECTROMETRY

As indicated in Chapter 3 (Section III), a mass spectrometer is an
instrument which sorts out charged gaseous species (molecular ions
and fragments) according to their masses. The positively charged ions
are produced by bombarding molecules of the gas or vapor to be analyzed with rapidly moving electrons. The energy of the ionizing beam
is controlled by the magnitude of the potential difference used to accelerate the electrons.

The mass spectrum obtained using a minimum energy beam (ca.
10 eV) consists almost entirely of a single peak, corresponding to the
mass of the original molecule (i.e., $ABC + e \rightarrow ABC^+ + 2e$). However,
for most applications the electron beam is given an energy of 50 to 100
eV. This produces excited charged species which fragment to yield a
mass spectrum composed of some parent peak plus a characteristic
fragmentation pattern.

Since collision with other molecules or fragments can vitiate the
separation of species on the basis of their relative masses, mass
spectrometers operate under a high vacuum (e.g., $< 10^{-9}$ torr), use
very small samples, and the ions are pulled out of the ionizing chamber
as they are formed by means of negatively charged accelerating electrodes (Fig. 3.8).

The potential differences across the accelerator grids are of the
order of 1 to 100 kV, and during their acceleration the charged species
acquire energy (F) equivalent to eV where e is the charge on the ion
and V is the applied potential.

The acquired energy (eV) is converted to kinetic energy ($mv^2/2$),
hence by equation of these terms, velocity $v = (2eV/m)^{\frac{1}{2}}$.

Besides accelerating the ions, the electric field assists in focusing
them into a fine beam which can be fed into a strong magnetic field at
normal incidence to the boundary of the field.

In magnetic fields (of flux H oersted), charged ions are subject to
a magnetic force given by $F = Hev$ which is equaled by a centrifugal
force given by $F = mv^2/r$ (where r is the radius of a circular path).
By equating the two terms and substituting for v one obtains the relationship $r = (2Vm/e)^{\frac{1}{2}}/H$.

From this equation it follows that the trajectories of ions of different mass differ, and if one has a fixed detector, the mass range in a
sample may be scanned by maintaining H constant with progressive
variation of V; or the reverse procedure may be used.

Figure 3.8 shows the arrangement of a typical mass spectrometer with magnetic scanning.

Mass spectrometers differ in geometric design, mode of scanning, and resolving power. The resolution of an instrument is a measure of its ability to separate ions of adjacent mass, and is defined as $M/\Delta M$, where M is the nominal mass of the pair of closely spaced peaks separated by ΔM mass units. It is desirable to achieve resolutions of 1 part in 10,000 or more so that the instrument can separate peaks such as N_2^+ (mass 28.0061) and CO^+ (mass 27.9949). With the aid of the spectrum of a marker compound (e.g., perfluoro kerosine may be admitted with the sample vapor) the measurements of the mass of the parent peak can be sufficiently accurate to allow absolute formula assignment.

The analytic applications of mass spectrometry are mainly restricted to gases and vapors. Gases are introduced through a calibrated leak to the ionizing chamber. The pressure required is about 10^{-3} torr and by using a heated sample introduction system (e.g., up to 250°C) most organic molecules yield sufficient vapors for study. The relative amounts of the parent ion and the numerous fragments produced by organic materials are characteristic of the molecule and can be used as a means of identification and for quantitative analysis. Identification is generally based on comparison of the unknown spectrum with standard spectra.

With a mixture of compounds the observed spectrum is the sum of the individual spectra, and if the behavior of pure components is known, one can set up a series of simultaneous equations which permit complete analysis of the mixture. The latter task has been simplified by the use of data processing equipment, but it is now considered preferable to pre-separate the mixture on a gas liquid chromatograph coupled directly to the mass spectrometer. The carrier gas (usually He) is removed by differential pumping through a porous tube as it emerges from the separation column, and the separated components yield their characteristic mass spectrum as they successively enter the ionization chamber.

It is possible to purchase mass spectrometers designed particularly for analytic purposes, for example, small radio frequency mass spectrometers have been carried by rockets and satellites and used for the analysis of simple gases in the upper atmosphere. Unfortunately, all are expensive and most are limited in application by the low volatility of many materials and complications attributable to fragment interactions. On the credit side, only very small amounts of material are required and they do provide a most precise means of analyzing gases and vapors. Extension to studies of nonvolatile solids requires alternative methods of ion formation. In some cases ionization is produced at the surface of a field emitter; in others, arc and spark sources produce the desired ions.

V. NEUTRON ACTIVATION ANALYSIS

At least 70 elements can become radioactive when bombarded with neutrons possessing thermal velocities. Capture of neutrons enlarges the atomic nuclei and causes instability which is manifested by spontaneous decomposition and the emission of a particle or a γ ray:

$$_{11}^{23}Na + _{0}^{1}n \rightarrow _{11}^{24}Na + \gamma$$

The active isotopes formed in this way differ widely in half-life values[*] and in many instances can be identified by determination of this constant, along with other pertinent information, such as the mode of decay, energy of emissions, etc.

In a mixture of independently decaying radionuclides, the total activity A at time t is given by

$$A = A_o' \exp \frac{-0.693t}{T'} + A_o'' \exp \frac{-0.693t}{T''} + \cdots$$

where A_o' and A_o'' represent the initial activities of separate species having half lives of T' and T'' respectively.

Provided the half-lives are sufficiently different, the decay of the mixture can be resolved graphically (Fig. 6.6) or analytically by solution of simultaneous equations.

The activity produced by a given element can be calculated from the equation

$$A = N\sigma f[1 - \exp(-\lambda t)]$$

where A is the observed radioactivity (disintegrations/sec); N is the number of target atoms (= wN_o/M where w is the weight of element of atomic weight M present, and N_o is Avogradro's number); σ is the atomic activation cross section for the nuclear particle reaction

---

[*] The half-life is the time required for any given sample of an isotope to be reduced to one-half of its initial value. The rate of decay is proportional to the number (N) of excited atoms present, i.e., $-dN/dt = \lambda N$ where λ is the decay constant which can be shown to equal $0.693/T$, T being the half-life.

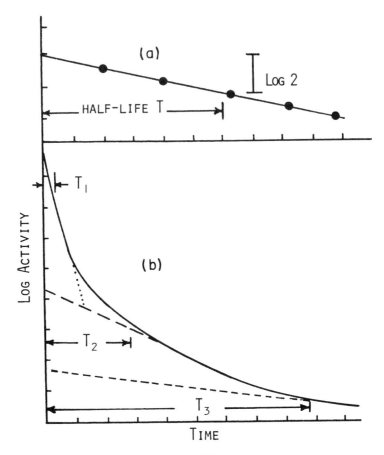

Fig. 6.6 Radioactive decay. Plot of log activity versus time: (a) single radioactive species, and (b) mixture of three radiation emitters. (Reproduced from Pickering, W. F.: "Modern Analytical Chemistry," Dekker, 1971 with permission of the publisher.)

(10^{-24} cm^2) f is the nuclear particle flux (particles cm^{-2} sec^{-2}), and λ is the decay constant. The term 1 - exp (-λt) is sometimes called the growth factor. It has a value of 0.5 when the time of irradiation equals the half-life of the nuclide, and approaches its maximum value of unity when the irradiation time is >6 half-lives.

In neutron activation analysis the sample is exposed to a high neutron flux, and on removal from the source, the activity is monitored as a function of time. The measuring equipment used can be based on evaluation of the degree of ionization created in a gas tube (Geiger tubes or proportional counters); or on measurement of the amount of visible light produced on a phosphor. In addition, the groups of pulses may be separated on the basis of their energy content.

Since detectors are not fully efficient, for comparative results a further term ψ needs to be added to the above equation; ψ indicates the efficiency of the detector system.

It can be seen from the preceding equation that the sensitivity of neutron activation analysis depends upon the intensity of the source, the ability of the element to capture neutrons (σ), the atomic weight of the element, the half-life of the nuclide formed, and the character of the radiation emitted, since this influences detector efficiency.

The flux in a nuclear pile can be of the order of 10^{13} n cm^{-2} sec^{-1} and this is sufficient to activate nearly all elements heavier than neon, but with greatly differing sensitivities. As little as 100 pg of some elements can be detected. A ^{124}Sb/Be source which produces a flux of 10^3 to 10^4 cm^{-2} sec^{-1} can activate about twenty elements.

The weight of a given element present in a sample is usually determined by a comparison technique. A known amount of the element of interest is bombarded simultaneously with the test sample and is processed, after its activation, in the same manner as the test sample [W(test)/W(comparator) = A(test)/A(comparator)]. Where more than one element is activated, the components have to be identified from the decay curve (Fig. 6.6), or be isolated by chemical separation procedures, or be separated on the basis of the energy content of the different emissions.

Apart from the usual precautions required in handling radioactive material, activation analysis is subject to a number of limitations including self-absorption effects and sample destruction.

The major fields of application include:

1. Determination of elements present at <0.001% level. In this range, technique is more sensitive than other conventional methods.

2. Minor constituent analysis (0.001 to 1%). With species having half-lives of the order of minutes to hours, the methods can compete in cost and simplicity with other approaches.

3. Fast automatic analyses where chemical separation is avoided through the use of sophisticated instrumentation.

A wide variety of sample materials have been examined by this technique, and it has been used in many pollution evaluation studies (e.g., examination of suspended particles from the air, analysis of hair and plant materials, determination of the mercury content of fish, etc.). Assay samples come from all branches of science, and the scope is being greatly enhanced by the increasing availability of neutron generators designed specifically for activation analysis.

Isotopes produced by an activation process can be used as tracers in chemical reactions or as tracers in pollution studies.

Much simpler measuring equipment is required in radioactive tracer techniques, since one is concerned primarily with changes in the distribution of the radiation derived from one particular source.

The behavior of a radioactive isotope of an element in chemical or physical processes is identical with that of its stable isotopes. Thus if a small amount of active isotope is mixed with an excess of the more stable forms, it can be used to trace the behavior of the bulk material through a whole series of simple or complex operations.

If the specific radioactivity remains virtually constant throughout the period of the experiment, then the radioactivity observed after some chemical process, or distribution, is directly proportional to the quantity of the element involved. The procedure can thus be used to observe adsorption processes or the extent of co-precipitation, or to determine distribution coefficients, etc.

In pollution studies, tagged material can assist in identifying sources of contamination, can aid understanding of species distribution in multiphase systems, can facilitate studies of losses due to sorption phenomena, be used to test the efficiency of chemical or physical separations, and so on.

7

SOIL, PLANTS, AND FOOD

I. NATURE OF SYSTEMS

Until recent years, analytical interest in natural species was confined
to evaluation of factors contributing to the fertility of soil or the nutri-
tional value of plants and foods.

   With the recognition that trace inorganic elements and pesticides
are also important factors in toxicology and environmental pollution,
equal emphasis is now placed on the determination of contaminants such
as lead, cadmium, mercury, arsenic, selenium, and organic residues.

   The analytical procedures used to evaluate these pollutants utilize
most of the techniques described in preceding chapters. Thus toxic
metals may be determined by means of colorimetric measurements,
polarography, or atomic absorption spectroscopy. Pesticide and
herbicide residues, after isolation by extraction, may be quantitatively
identified by infrared studies, gas chromatography, or mass spectrometry.

   The evaluation stage, using the appropriate technique, is usually
quite straightforward but determinations are often beset with uncertainties associated with the preliminary stages, i.e., sampling and component isolation.

   The uncertainties arise, in part, from considerations such as the
following.

   1. Natural systems are distinctly heterogeneous. In plants,
      one can obtain different values in studies of roots, nodules,
      stems, and leaves; metals are concentrated in the germ
      or bran fractions of many edible grains; and nutrients in
      soils can be preferentially sorbed by particular components.

2. <u>Distinction should be made between active forms and total content</u>. It has long been the practice to subdivide the total phosphorus content of a soil into organic phosphorus compounds and inorganic. The latter are often further divided into the categories of P from aluminum, iron and calcium phosphates, and P occluded in these specific compounds. Fractionation is achieved through successive treatments with chemical solutions such as 1 M NH_4Cl; 0.5 M NH_4F; 0.1 M NaOH; 0.25 M H_2SO_4; Na-citrate, Na-dithionate; etc. To ascertain available pollutant species similar extraction procedures need to be adopted. Unfortunately the efficiency of an extractant process can be difficult to assess and there can be debate about the nature of the active form.

3. <u>Processing can alter component concentrations</u>. Only certain parts of a plant may be eaten, and components may be lost (or enhanced) during refining processes or preparation for consumption. For example, vegetables lose many trace metals during cooking. Hence, should analyses of these species be preceded by a standard cooking process?

4. <u>Responses to environmental changes are variable.</u> There is no simple relationship between soil levels for a given element, vegetation levels, and content in other species along the food chain. Plants derive some contaminant content by uptake from the soil and some through sorption of vapors and particulates from the surrounding air. Fauna feeding on the flora retain varying proportions of the total intake. In marine environments, concentration effects can increase the contaminant levels to dangerously high values (Table 7.1). Land plants can also act as concentration units for specific elements.

II. SAMPLING AND SAMPLE PREPARATION

Because of the distinct heterogeneity of most natural systems, the usefulness of any results can depend entirely on the care taken during sampling. At the same time, selection of an appropriate sampling procedure is not necessarily a simple task.

For example, one book on soil analysis begins its chapter on soil sampling with the quote: "In view of the variability of soils, it seems impossible to devise an entirely satisfactory method for sampling. It is obvious that the details of the procedure should be determined by

Table 7.1

Trace Metal Content of Marine Species[a]

Sample	Copper		Lead		Cadmium		Zinc	
	Conc.[b]	BCF[c]	Conc.	BCF	Conc.	BCF	Conc.	BCF
Seaweed	0.4	100	0.7	500	0.8	4000	17	4000
Fish flesh	2.0	400	0.4	300	0.04	100	73	2×10^4
Abalone	2.3	400	0.1	100	0.04	200	41	9000
Oysters	13	3000	0.9	600	4.4	2×10^4	2700	6×10^5

[a] Data taken from Florence, T. M.: J. Electroanal. Chem. 35:237, 1972.

[b] Concentration in dried sample, ppm.

[c] Biological concentration factor (ppm in dried sample/μg ml^{-1} in seawater.

the purpose for which the sample is taken" (AOAC, Official and Tentative Methods for Analysis, 6th ed., Washington, 1945).

Similar comments can be made about sampling marine biota, crops, vegetation, etc. Even processed foods are not homogeneous in quality, and it is desirable to collect and adequately blend a number of subsamples before assessing the quality of a given batch.

Where the quality of a product is defined by legal or contract limits, an established sampling procedure is usually included in the approved, or official, analytical methods. In new studies, sampling procedures need to be justified by variance studies (Chapter 2). Some check is also required on errors introduced during blending and reduction of gross samples to laboratory size, since there can be loss of moisture and other volatile components, changes induced by microbiological activity, or general contamination. Even storage can be a problem since volatile components can be lost to plastic containers, or conversely, components of the container (e.g., vinyl chloride) may diffuse into the sample.

If one assumes that the laboratory sample is truly representative, the second stage is release of the species of interest for quantitative study.

Table 7.2

Modes of Dissolution for Organic Compounds

Type	Descriptions	Applications or comments
Dry ashing	Air oxidation in crucible, flame, or furnace	Inorganic species in residue. Some Hg, Cd, Cu, As, and Ag may be lost
	Hot-tube furnace, oxygen stream	Gaseous products collected for determination of major elements (C, H, N).
	Low-temp. oxidation with "active" O_2	Low oxygen pressure, used for residue analysis, volatile products can be trapped.
	Combustion in (1) oxygen flask;	Products sorbed in liquid; used for determination of S, halogens, and metals.
	(2) oxygen bomb	High temperature and pressure (e.g., 20 atm); determination of S and halogens.
Wet ashing	Oxidizing acids (1) Single, HNO_3	High temp. useful for biologic samples.
	(2) Mixed, $HNO_3 + HClO_4$	Protein and carbohydrate determinations.
	$HNO_3 + H_2SO_4$	Plants.
	$HClO_4 + HNO_3 + H_2SO_4$	Universally applicable and is preferred method; allows rapid oxidations of difficult samples.
Peroxide attack	$H_2O_2-H_2SO_4$	Powerful; useful for organometallics, plants.
	$H_2O_2-HNO_3$	Biologic samples, rapid for small samples, low temperature.

Table 7.2 (Continued)

Type	Descriptions	Applications or comments
	H_2O_2-Fe^{2+}	Ashing occurs through OH- radical formation, and at about $100^{\circ}C$. Widely used (except fats, oils, plastics).
	Peroxide bomb (Parr)	Sample burnt in Na_2O_2 containing accelerator (KNO_3, $KClO_3$ or $KClO_4$); used for determinations of S, halogens, P, B, Si, As, and Se.
Fusion	Alkali metal in sealed tube	Determination of halogens.

Some materials can be analyzed directly after simple dissolution in water or other solvent, but most samples require one or more of the following three basic treatments.

1. Acid dissolution and hydrolysis. On boiling samples with acid (e.g. 6 M HCl or HNO_3 for 10 min inorganic particulates tend to dissolve and proteins are hydrolyzed. After removal of residual solids by centrifugation, the solution should contain most of the initial trace metal content (carbohydrate and fat fractions are generally not destroyed or dissolved).

2. Extractive leaching. Many chelating agents are available which could, in principle, extract the element of interest without complete hydrolysis or destruction of the sample. For example, boiling with 0.1 M EDTA at pH 6 can lead to extraction of Cu, Ni, Fe, and Cr from many samples.

3. Complete sample dissolution and destruction of organic matter. The two usual procedures for treating soil samples are fusion in Na_2CO_3 or heating with acid mixtures (e.g., HF, $HClO_4$, H_3PO_4, H_2SO_4). For the destruction of organic compounds, wet and dry methods are available, as summarized in Table 7.2. The choice of technique is dictated by the type of material being examined and the type of analysis required.

III. SEPARATION PROCEDURES

After release of the components of interest, it is often necessary to insert another basic procedure to eliminate residual matrix effects. The clean-up process can involve bulk removal of interferants or selective isolation of the species of interest.

For some toxic metals (e.g. As, Se, Hg) special vapor generation techniques have been developed. In arsenic and selenium determinations, for example, the digested sample is treated with zinc powder in an acidified stannous chloride-potassium iodide matrix. The gases evolved (AsH_3, H_2Se) are fed into the flame segment of an atomic absorption spectroscopy unit. In the flame the gases are atomized and the peak absorption is noted on a recorder.

For the isolation and preconcentration of other elements extraction into an organic solvent is widely used.

When solutes are added to any system composed of two immiscible liquids, they distribute themselves between the two phases. Charged ions (i.e., polar species) tend to remain in the more polar solvents (e.g., water), while uncharged (covalent) species prefer an organic solvent phase.

To cause a metal ion to transfer quantitatively to an organic solvent, it is thus necessary to mask its polarity by converting it into a coordination complex. Organic chelating agents, such as 8-hydroxyquinoline, are very useful for this purpose.

Consider the extraction of aluminum into chloroform as aluminum 8-hydroxyquinolate.

The chemical equilibria involved in this process may be expressed in the following manner.

Aqueous Phase   Organic Phase

$$H^+ + Ox^- \underset{K_a}{\overset{K_a}{\rightleftharpoons}} (HOx)_{(aq)} \underset{K_D}{\overset{K_D}{\rightleftharpoons}} (HOx)_{(o)}$$

$$Al^{3+} + 3Ox^- \overset{K_c}{\rightleftharpoons} Al(Ox)_{3\,(s)} \overset{K_E}{\rightleftharpoons} Al(Ox)_{3\,(o)}$$

It can be observed that the amount of metal in the organic phase is determined by the values of K_a, K_D, K_c, and K_E, and the pH, since this controls the amount of ligand (Ox^-) available to form the complex.

Since K_c varies in magnitude with the metal ion involved, it is often possible to select conditions whereby one species is almost wholly extracted while other species remain in the aqueous phase. Selectivity can be enhanced by using organic reagents which react specifically with a limited number of metal ions or by adding complexing agents which form stable ionic species with interfering ions (i.e., masking).

For a given extraction process, a known volume of organic phase (containing the selected reagent) is shaken with aqueous solutions of the metal ion of differing pH. Plots of percent extraction versus pH permit elucidation of optimum conditions for separation and quantitative recovery. By using a volume of organic phase which is smaller than the volume of the aqueous sample, preconcentration is simultaneously achieved.

Where the extracted material is colored, it may be used for colorimetric determinations; alternatively it may be fed into an atomic absorption unit.

If more than one species is extracted, the measuring technique may be modified to give the desired selectivity (e.g., different source lamps in atomic absorption spectroscopy).

The extraction of pesticide residues from samples by means of organic solvents may be regarded as a special case of solvent extraction. The extracts generally contain many large unwanted molecules, and these may be removed by passage through a bed of adsorbent (e.g., activated charcoal) but final separation of isolated species is usually achieved by gas chromatography.

With this technique, a further degree of selectivity can be gained through the use of different detector systems. Thus electron-capture detectors are ideal for studying organochlorine pesticides, alkali flame ionization detectors respond preferentially to organophosphorus compounds, and so on.

IV. ILLUSTRATIVE METHODS OF ANALYSIS

Because of the diverse nature of soils, plants, and foods, variations in analytical procedure are manifold. However, the purpose of this introduction to pollution evaluation can be served by restricting the discussion to a few examples. The five methods considered in this section are thus intended to be indicative of general approaches, not definitive.

A. Determination of Copper (and Zn) in Soil

The total copper and zinc content of soil normally falls in the range of 5 to 40 ppm.

Decomposition of a soil through treatment with HF (in the presence of H_2SO_4) releases all the metal ion (including that bound in silicates). After twice evaporating off added HF, organic matter in the residue is destroyed by adding nitric acid and evaporating to dryness. The remaining salts are digested with a ternary acid mixture (HNO_3, H_2SO_4, $HClO_4$) and the total copper (or zinc) content is then determined either colorimetrically, polarographically, or by atomic absorption spectroscopy.

For the purpose of distinguishing between the various forms of copper present, several variations in procedure have been proposed. For example, the concentration of copper found after digestion of the soil with a $HClO_4$, H_2SO_4 mixture is considered to be a measure of the potentially available copper in organic soils. Other workers recommend the use of selective extraction agents for the identification of active modes, but it should be noted that the amount of soil copper extracted varies from nearly all of it (e.g., using $HClO4$) to a very small fraction (e.g., using neutral ammonium acetate). In a recent study (J. Soil Sci. 24:172, 1973) the copper content of soils was fractionated into five categories by using different extractants, namely, available and exchangeable Cu (0.05 M $CaCl_2$ treatment); weakly bound Cu (2.5% acetic acid digestion); organically bound Cu (released with 1 M $Na_4P_2O_7$); occluded by oxides (oxalic acid, pH 3.25); and lattice copper (released with HF).

In the case of zinc, an estimate of the fraction available to plants has been obtained by shaking the soil with a two-phase system composed of 25 ml of 1 M ammonium acetate (pH 7) and 25 ml of carbon tetrachloride containing the organic reagent, dithizone. The zinc extracted is retrieved from the organic solvent by back extraction into dilute acid, and is subsequently determined by standard methods.

B. Determination of Lead in Leaves

Reported values for the lead content of leaves and other vegetation usually fall in the range of 10 to 100 $\mu g/g$. In most areas, the prime source of this element is fallout from motor vehicle exhausts, hence levels vary with distance from highway, time of exposure, height above ground, etc. Washing the leaves (e.g., with detergent solution) before analysis can reduce observed values by half. To obtain reasonable average values, it is thus necessary to collect, dry, and thoroughly blend a large number of leaves before weighing out the analytic sample.

The organic content of the leaves is readily destroyed by digesting the sample with a mixture of nitric and perchloric acid, followed by evaporation to fumes of perchloric acid. After dissolution of the cooled residue in dilute nitric acid, the lead content can be determined by atomic absorption spectroscopy, or absorptimetry, or polarography.

Where preconcentration is desirable because of low levels, the lead may be extracted at pH 2.5 into methyl isobutyl ketone as the lead-APDC complex (APDC = ammonium pyrrolidine dithiocarbamate). The extract can be fed directly in an atomic absorption spectroscopic unit, using either atomization by flame or carbon rod.

Punched disks (2 mm in diameter) from individual leaves have been examined in a carbon cup attachment (particulate Pb, Chapter 3), but this serves primarily to show the variation in Pb distribution within a given leaf (e.g., in eucalyptus leaves the Pb appears to be most concentrated at the edges).

C. Cadmium in Food

Cadmium in foods may arise from dust fallout on soil or from the use of contaminated superphosphate fertilizers, as well as from natural sources. It is an extremely toxic element, hence the levels present should be much less than a μg/g (levels of 3 to 4 μg/g net weight have been reported for oysters).

Cd is lost in dry-ashing procedures, hence a wet digestion method using H_2SO_4 and nitric acid is recommended. Digestion is continued until oxidation is complete and fumes of SO_3 are being evolved. The cooled digest is diluted with water, filtered, and citric acid added before introducing ammonia to adjust the pH to about 9.

Reactive metals are separated from this solution by shaking with a solution of dithizone in chloroform. The cadmium is then separated from any Cu, Hg, Ni, or Co present by stripping the chloroform solution with dilute HCl.

The acid solution (containing the Cd) is adjusted to 5% NaOH, before being shaken with a dithizone-carbon tetrachloride solution. At this alkalinity Zn, Pb, and Bi do not extract, whereas Cd dithizonate is relatively stable. The Cd is finally estimated photometrically as the colored dithizonate. (Zinc constitutes the chief interference.)

Alternatively the initial digest may be examined in an atomic absorption unit, either directly (detection limit ca. 0.001 μg/ml) or after extraction of Cd-APDC complex at pH 1 to 6.

D. Mercury in Fish

Mercury has been shown to accumulate in the food chain of fish, and large predator fish such as tuna and shark may accumulate several $\mu g/g$, mainly as methyl mercury compounds. Most countries impose legal restrictions on the permissible levels; typically, 0.5 $\mu g/g$ has been set as the limit.

Wet ashing is essential for sample preparation, and to minimize losses oxidation conditions must be maintained. One approach adds 0.5-g increments of the total sample to hot nitric acid, boiling after each addition until reaction subsides. Alternatively the sample can be heated gently, under a reflux condenser, with a mixture of sulfuric and nitric acids in the presence of a catalyst (molybdate). After an hour, the solution is cooled, a nitric-perchloric acid mixture is added, and the mixture boiled vigorously until white fumes appear. Another modification dissolves the sample in warm H_2SO_4 (takes 2 to 3 hr). These sample solutions are then cooled in ice before slow addition of some potassium permanganate solution. Oxidation is completed by standing overnight or by heating on a water bath for an hour.

Extraction of the mercury (at pH 2 to 4) into chloroform containing dithizone, and subsequent colorimetric determination was the recommended evaluation procedure until the more sensitive cold-vapor atomic absorption techniques were developed.

In the cold-vapor technique, aliquots of the sample digest are placed in a closed test tube with some stannous chloride, and the mixture is stirred vigorously for a fixed period of time (e.g., 90 sec). This saturates the air in the tube with mercury vapor. The stirrer is turned off and a stream of air is passed through the tube to convey the mercury vapor into a quartz cell mounted in the light beam (Hg lamp) of an atomic absorption unit. The absorption peak recorded is related to mercury content with the aid of standards. The detection limit is about 2 ng of Hg.

To identify and determine the methyl mercury compounds in fish, it is necessary to use mass spectrometry, or a combined gas chromatographic, thin-layer chromatographic method.

E. Determination of DDT (Dichlorodiphenyltrichloroethane)

Before scientists became concerned with the build up of DDT in the environment, this was the most widely used pesticide, being applied either in the form of dusting mixtures or sprays.

In the absence of other organic chlorine compounds, the standard method of analysis is that based on the determination of the chlorine content.

Thus in studies involving technical-grade DDT, samples containing about 75 mg of pesticide are dissolved (if necessary) in a small volume of benzene before addition of isopropanol (e.g., 25 ml). A few grams of sodium metal are mixed with the sample and the flask is boiled gently for about half an hour under a reflux condenser before excess sodium is destroyed by cautious addition of more isopropanol.

After disconnection of the condenser, about 60 ml of water are added and the solution is boiled to expel residual alcohol. The chloride content of the aqueous solution is finally determined by titration with a standard solution of silver nitrate.

In the presence of organic matter (coloring matter, plant resins, mineral oils, etc.) the initial solution in benzene is followed by addition of some decolorizing carbon, and filtration. Residual organic contamination is then removed by adding an excess of dilute nitric acid, followed by shaking with a volume of an isoamyl alcohol-ether (1:1) mixture. After separation of the phases, the organic layer is washed with water, and the chloride content of the combined aqueous layers determined.

An alternative approach to DDT analysis uses infrared absorption as the measuring technique. The sample is dissolved in carbon disulfide and the solution dried by adding anhydrous Na_2SO_4. A portion is then transferred to a NaCl cell and the infrared sorption measured in an appropriate spectrometer over the wavelength range 8.5 to 10.5 μm. Standard solutions are scanned in the same manner and the intensity of the 9.83 μm peak is used for comparison purposes.

These methods are not completely selective, and in mixtures, quantitative analysis has to be preceded by a separation procedure.

Gas-liquid chromatography (using an electron-capture detector) is the most commonly used method for quantitative <u>pesticide residue analysis</u>, but it is subject to misidentifications and errors as a result of interferences from coextracted pesticide compounds and naturally occurring products.

One useful technique for confirmation of residue identity is chemical derivatization. Through selective reactions, the pesticides or metabolites concerned are converted (before injection into a chromatograph) into derivatives with different retention times from the parent compounds and also from other common pesticides that may be present.

Of further value is information derived from the response behavior observed when using several of the more or less specific gas chromatographic detectors currently available. Spectrometric techniques (e.g., infrared and ultraviolet spectrophotometry) provide less ambiguous proof of identity and the value of mass spectrometry as a tool for providing the structural identity of complex organic molecules is well recognized.

If evidence from more sophisticated techniques is not available, adequate criteria of identification may be derived from paper or thin-layer chromatography studies.

A small spot of cleaned sample mixture is placed near one end of a sheet of filter paper or a glass plate coated with a fine layer (e.g., 100 μm) of adsorbent material (cellulose, alumina, silica gel).

Spots of pure, suspected components are placed along the same horizontal line. The end of the sheet or thin-layer plate is then placed in a reservoir of appropriate organic solvent mixture and the system enclosed by a glass vessel. The solvent mixture migrates up the solid supports by capillary action, and if the nature of the support and solvent mixture has been correctly chosen, then the components of the mixture are separated into a series of spots (as shown in Fig. 7.1). The ratio

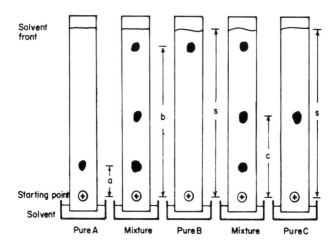

Fig. 7.1   Diagram showing the separation of a three component mixture on a thin-layer chromatogram. The R_f values of suspected components A, B, and C have been identified from the comparison chromatograms of pure compounds; the R_f of A is a/s, the R_f of B is b/s, etc. (Reproduced from Pickering, W. F.: "Modern Analytical Chemistry," Dekker, 1971 with permission of the publisher.)

of the distance moved by a given species, to the distance moved by the solvent is known as the R_f value. A similarity in R_f values is circumstantial evidence of similarity in identity between sample components and standards. In order to determine the R_f values, it is necessary to visualize the various species through treatment of the supports with suitable chromogenic reagents.

FURTHER READING

Horwitz, W., ed.: "Official Methods of Analysis of the Association of Official Analytical Chemists," 11th ed., A.O.A.C., Washington, 1970.

Jackson, M. L.: "Soil Chemical Analysis," Constable, London, 1962.

"Pesticides Identification at the Residue Level," Advances in Chemistry Series 104, American Chemical Society, Washington, 1971.

Zweig, G., ed.: "Analytical Methods for Pesticides, Plant Growth Regulators and Food Additives," Vols. I-V, Academic Press, N.Y., 1963-7.

PRINCIPLES OF PRECONCENTRATION, MASKING, AND METHOD SELECTION

In the brief treatments of analytical techniques presented in preceding chapters, two important aspects have received limited comment.

Firstly, while the detection limits of many procedures are of the order of mg/liter or less, the percentage relative error is extremely high at these limits, and interference effects are maximal. Accordingly, many advantages can accrue from adopting some preconcentration procedure. The larger amount of material studied (e.g., increased by a factor of >10) minimizes reading errors, a change of media (e.g., into an organic solvent) may enhance the instrument response, and in favorable situations elimination of matrix problems may be achieved (e.g., by selective isolation).

The second omission in previous discussions has been direct comment on modes of eliminating interference effects. In most analytical determinations, a significant proportion of the laboratory procedure is designed to eliminate or minimize interference effects.

The general approaches to interference control fall into four main classifications.

1. Refinement of instrumental technique. Examples are background correction in atomic absorption spectroscopy, mixed color methods in absorption spectroscopy, cathode potential control in electrogravimetry.

2. Selective isolation of the species of interest. The techniques used include precipitation, solvent extraction, volatilization, controlled oxidation or reduction, adsorption on ion-exchange materials, and chromatographic methods.

3. Removal of interferants. The techniques used are basic
ally similar to 2, but as the mass of material handled is
larger, closer control can be needed to prevent loss of the
desired species, e.g., by co-precipitation, co-extraction,
etc.

4. Masking. The characteristic reaction of an interferant
species with a given reagent is masked through formation
of a new stable chemical form, e.g., the addition of
cyanide ions can be used to prevent the precipitation of
copper and cadmium sulfides on the addition of hydrogen
sulfide to a solution containing these ions.

In this chapter, the fundamental principles of some of the more
widely adopted preconcentration or separation procedures are considered briefly.

I. PRECONCENTRATION PROCEDURES

A. Solvent Extraction

Extraction of a sample component into a small volume of organic solvent
can be highly selective, and may serve to both concentrate and enhance
technique sensitivity. The isolated material may be fed directly in the
measuring technique (e.g., radiation absorption studies) or be returned
to the aqueous phase for further treatment.

In solvent extraction, an aqueous solution containing a mixture of
chemical species is shaken in a separating funnel with a volume of
water-immiscible organic solvent. "Like dissolves like," hence nonpolar components tend to transfer to the organic phase, leaving polar
compounds such as metal salts in the aqueous phase. The art and
science of this approach lies in being able to adjust conditions in the
aqueous phase so that only one species is extracted. The species should
be quantitatively transferred to the extracting phase; and as noted in
later sections, the efficiency of the transfer process can depend on
several factors, including solution pH, conversion to extractable forms,
type of solvent, etc.

The types of compounds which are solvent extracted fall into the
following distinct groupings.

Organic and inorganic covalent compounds. Examples are the re-
trieval of trace amounts of fats, oils, hydrocarbons from natural waters;
or the extraction of the halogens, Cl_2, Br_2, I_2.

Metal chelates. The formation of a metal chelate yields an uncharged compound which is soluble in the organic solvent, (Section V. A of Chapter 5).

Ion-association compounds. The polarity of inorganic ions can be masked through coordination with organic groups to provide a large ion which is paired in the solvent with an ion of opposite charge. Alternatively, the ion of interest is the partner of a large organic ion of opposite charge. For example,

$$M^{n+} + x(Org) + nA^- \rightleftarrows M(Org)_x^{n+}, nA^- \text{ (extracted)}$$

$$M^{n+} + y\ L^- + (y-n)Org^+ \rightleftarrows ML_y^{(y-n)}, (y-n)Org^+ \text{ (extracted)}$$

Solvent adducts. Organic solvent molecules which occupy coordination sites give the molecular species or ion pairs affinity with the extraction medium. For example, iron(III) is extracted into ether from HCl solutions probably as $[(C_2H_5)_2OH^+, FeCl_4 \{C_2H_5)_2O\}_2^-]$. In the presence of tributyl phosphate, uranium can be extracted as $UO_2(NO_3)_2 2Bu_3PO_4$.

Liquid ion exchangers. Ions from an aqueous phase can be caused to replace charged species attached to exchanger materials present in an organic phase. For example, the extraction of the uranyl ion can be represented as

$$UO_2^{2+}_{\text{(aq)}} + 2(HX)_{2\text{(org)}} \rightleftarrows UO_2 \cdot 2HX_{\text{(org)}} + 2H^+_{\text{(aq)}}$$
$$\text{(exchanger)}$$

The exchanger materials, often organic esters of phosphoric acid (e.g., heptadecyldihydrogen phosphate), are used as 5 to 10% solutions in solvents such as benzene, kerosine, carbon tetrachloride, or cyclohexane.

The term of most interest in any type of system is the distribution ratio D (= total concentration of species in organic phase/total concentration of species left in aqueous phase).

In most extraction procedures the efficiency of the process (i.e., size of D values) varies with the composition of the aqueous phase. Accordingly, solvent extraction procedures usually define the optimum pH for extraction, the masking agent required (if any) to prevent extraction of other species, the type and concentration of reagent required to form extractable species, and recommended organic phases.

For some systems it is possible to calculate optimum extraction conditions from available equilibrium data, but for most applications it is necessary to plot experimental extraction curves. The experimental curve is prepared by determining the percentage of original metal ion transferred when one of the solution variables (e.g., ligand concentration or pH) is altered systematically (Fig. 8.1).

Solvent extraction is used mainly as a means of isolating one particular compound or group of compounds, and the conditions for effective separation (i.e., D > 100 or > 99% extraction with an equal volume of solvent) are best determined by examination of series of extraction curves. For example, from Figure 8.1b, it can be concluded that gold could be separated from iron(III) using 2 M HCl solutions and ether as solvent.

In selecting a new procedure it is essential to consider the kinetics of the extraction as well as the position of equilibrium. For many separations, simple repeated inversion of a separating funnel containing

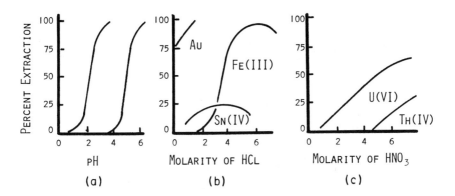

Fig. 8.1 Extraction curves. (a) The amount of metal chelate extracted into an organic solvent is a function of both pH and complex stability (system 1 > system 2). (b) and (c). The efficiency of extraction of ion-association systems is less predictable. The curves illustrate the effect of acid concentration on the extraction of some metal ions into diethyl ether. (Based on data from Bock, R., and Bock, F., Z.Anorg. Chem. 263:146, 1950.)

the two phases is sufficient to achieve equilibrium in a matter of minutes. On the other hand, the extraction of metal chelates (e.g., using dithizone or acetylacetone) can proceed at a slow rate, especially if low concentrations are involved. The rate of extraction can also be reduced by the presence of masking agents in the aqueous phase.

If the extraction process is selective, but less than 100% efficient, complete recovery of the material requires repeated extraction with fresh volumes of organic phase. (With a 50% efficient process, eight extractions are needed to give 99.9% recovery since each operation removes only half of the residue.)

Where more than one species is extracted in the initial operation, purification may be achieved by back extraction into an aqueous phase of different composition. Alternatively, the subsequent analytical procedure may be sufficiently selective to allow use of the mixture (e.g., in atomic absorption spectroscopy).

The degree of preconcentration achieved with solvent extraction is normally a factor of about 10 (i.e., the content of 100 ml aqueous phase is collected in 10 ml of organic phase), but if desired, the factor may be increased further by volatilizing the initial solvent and dissolving the residue in a smaller volume.

B. Ion Exchange

1. Exchange Affinity

A number of materials, both natural and synthetic, are capable of exchanging ions in their structure for ions in a surrounding aqueous solution. For example,

$$2Ex \cdot Na^+ + Ca^{2+} \rightleftarrows Ex_2Ca^{2+} + 2Na^+$$

$$2Ex \cdot Cl^- + SO_4^{2-} \rightleftarrows Ex_2SO_4^{2-} + 2Cl^-$$

In these equations, Ex represents a solid insoluble matrix to which is attached the functional groups responsible for the exchange properties.

The mostly widely used exchangers in analytical applications are synthetic organic polymers which incorporate large numbers of a single functional group, e.g., $-SO_3H$, $-COOH$, $-OH$, $-PO_3H_2$ (for cation exchange) or $-NH_2-$ $-NHR$, $-NR_2$, $-NR_3$ (for anion exchange). The total number of ionized functional groups per unit of resin volume determines the exchange capacity.

While the materials are virtually insoluble in water, in many respects they act like concentrated solutions of acids and bases, and the number of charged groups available for exchange reactions can be controlled largely by the pH of the surrounding solution (similar to ionization of acids). Resins containing strong acid groups (e.g., $-SO_3H$) readily release protons and take part in exchange over a wide pH range, but those possessing weak acid functional groups ($-COOH$, $-OH$) only tend to ionize completely under alkaline conditions.

To achieve a quantitative exchange of ions, dilute aqueous solutions of the species of interest are allowed to flow slowly (e.g., ml/min) through a column of exchanger material. The dimensions of the exchanger bed can be, say 20 cm long and 1 cm in diameter, and with synthetic resins, each cm^3 of material has a capacity of about 150 mg of ionic material. Before use the resin is charged in a desired ionic form (e.g., $-SO_3H^+$ or Na^+ or NH_4^+) by running several bed volumes of concentrated electrolyte solution through the column. Excess electrolyte is subsequently removed by passage of distilled water.

After an appropriate volume of test solution has been passed through the column (for collecting trace components several liters may be used) the collected material may be released by passage of a small volume (e.g., two bed volumes, approximately 50 ml) of concentrated electrolyte (e.g., 0.1 M acid) through the exchange bed. The concentration factor is equal to the ratio of the initial sample volume to the regeneration volume.

Due to slight differences in the affinity of various ions for the exchanger, it is possible to have some ions retained preferentially or to have one ion eluted from the column before another. Any inherent differences in affinity can sometimes be enhanced by the presence of a complexing agent in the mobile phase.

Views differ on the nature of the mechanism responsible for ion exchange, and on the factors influencing selectivity. Hence for practical applications a few empirical rules (such as given below) often prove to be better guides to behavior than theoretical treatments. With cation exchangers and dilute aqueous solutions, the tendency of hydrogen ions to exchange varies with the nature of the functional group ($-SO_3H >$ $-COOH > -OH$). For ions of the same valence the extent of exchange increases with increasing atomic number of the exchange ion (e.g., $Li^+ < Na^+ < K^+ < Rb^+ < Cs^+$). With ions of differing valency the extent of exchange increases with increasing charge (e.g., $Na^+ < Ca^{2+} < Al^{3+} < Th^{4+}$).

A sequence of affinities such as hydroxide $>$ sulfate $>$ nitrate $>$ phosphate $>$ chloride can be proposed for anion exchangers, but with these materials the actual order seems to depend greatly on the functional groups present.

2. Analytical Applications

The exchange of ions on a column is used in analytic procedures for concentrating dilute solutions, for determining the total salt content of aqueous solutions, and for removing interfering ions.

In concentration applications, large volumes of test solutions (e.g., river water) are passed through the exchange column. If the affinity of a given species for the exchanger is much greater than the affinity of other ions present, retention and concentration may be achieved in the presence of significant amounts of other ions.

Increased selectivity can be achieved in several ways. The most effective is to make use of complex formation, for example, species which form stable anionic complexes can be separated from a wide range of other cations through adsorption on an anion exchanger. In this case the sample must be treated with an excess of ligand to ensure that the anionic complex does not dissociate prior to the regeneration or release stage. Alternatively, selectivity may be based on the preference of an exchanger for highly charged ions, or one can choose a chelating resin which contains ligand-type functional groups as well as exchange sites.

Expressed in general terms, ion exchange is a highly convenient means of separating electrolytes from nonelectrolytes. Well-known applications include retention of calcium, magnesium, and heavy metal ions from natural waters; removal of strontium, calcium, and copper ions from milk; isolation of traces of lead, copper, and iron from wine; etc.

The efficiency of an isolation (i.e., preconcentration) process should always be checked using natural samples as well as distilled water standards. In natural systems, the elements of interest may be present as stable complexes, or adsorbed on solids; buffer salts (e.g., bicarbonate ions in sea water) can remove protons from weakly acidic functional groups. As a result the capacity of the column, pH of the eluant, and overall efficiency can vary with the volume of sample used.

The total electrolyte concentration of a dilute solution can be determined by causing all the cations (or anions) in the sample solution to be quantitatively exchanged for another ion, [e.g., hydroxonium (or hydroxyl)], which can be easily determined (e.g., by acid-base titration). This procedure is more convenient than evaporating large volumes of sample to dryness and weighing the residue, and is more informative if the result required is total salt content, rather than total solids (which would include colloids, such as clay, etc.).

The third application of interest is elimination of interference effects. In many experimental situations, the accurate determination of an anionic species is hindered by competing reactions attributable to various cations in the solution. Often this problem can be solved by first passing the sample through a cation exchanger charged in the hydrogen form. The replacement of metal ions by protons produces an equivalent amount of the corresponding acid in which one is free to quantitatively determine the anion content by any one of several alternative procedures.

Conversely, an anion exchanger charged in an appropriate form provides a convenient means of replacing anionic species which interfere in analytical procedures for specific cations. For example, phosphate ions in test solutions are regularly removed by passing the sample through an anion exchanger charged in the chloride form.

The separation of a mixture of ionic species through selective elution is known as ion-exchange chromatography. The basic principle is similar to that discussed in gas chromatography (Section II of Chapter 6). The mobile phase in this case, however, is an aqueous solution. The ionic strength may be varied to give sharper elution peaks or a complexing agent may be added to the mobile phase to shift the ion-exchange equilibrium in a direction which favors separation.

There are many modifications of this basic column procedure. The use of liquid ion exchangers in solvent extraction has been mentioned in section A; for the separation of biologic compounds, ion-exchange materials based on hydrophilic polymeric gels have proved extremely useful; and many inorganic separations and preconcentrations have been conveniently achieved using sheets of filter paper treated to increase the ion-exchange capacity of the cellulose.

Ion-exchange procedures can be adapted to handle microgram to gram quantities of material, and the technique has to be seriously considered whenever it is necessary to concentrate or separate ionic species.

C. Sorption by Solids

The adsorption of gases and solutes on solids can be used as a means of preconcentration and separation.

For discussion purposes, the adsorption of gases on solids can be considered to follow an isothermal relationship of the type,

$$\theta = \frac{Bp}{1 + Bp}$$

where θ is the fraction of the available adsorption sites covered, p is the pressure of the gas, and B is a constant whose magnitude is a function of the solid-gas interaction. (This simple Langmuir equation assumes all sorption sites to have equal activity and localized adsorption involving one molecule of gas per adsorption site.)

If the components of a gaseous mixture have different affinities for a solid (i.e., the values of the respective constants B vary significantly) then the species are adsorbed to different extents and selective removal of vapors from a gas stream is feasible. For example, combustion of organic compounds in a stream of oxygen produces a mixture of carbon dioxide, water vapor, and excess oxygen. The water vapor can be selectively retained on a column packed with solid desiccant (e.g., magnesium perchlorate) and the carbon dioxide subsequently held on an alkaline solid coating. The increase in weight of each sorbent is used to calculate the H and C content of the initial sample.

Where the adsorption of vapor leads to the formation of a colored species, the extent of the color change induced by the passage of a given volume of sample provides a semiquantitative measure of the concentration of reactive vapor present. Many of the standard routine procedures for detecting traces of toxic gases in the atmosphere are based on this principle.

Where the difference in affinity of various species is not so marked, repeated distributions between a mobile phase and the solid are required if a mixture is to be separated into individual components. This technique, known as gas-solid chromatography, has been discussed in Section II.B of Chapter 6.

The adsorption of solutes on to a solid surface has a variable role in analytical procedures. For example, sorption on the surface of a colloidal precipitate is detrimental if one is seeking to isolate and weigh the precipitate. On the other hand, if the aim is to quantitatively collect trace amounts of some ion, colloidal suspensions can serve as the carrier in the isolation process. For example, trace amounts of thorium in Zn, Cd, and Co salts have been isolated by using a precipitate of base metal arsenate to carry down the thorium content.

An adsorbent which is capable of collecting large amounts of organic and inorganic species from solution is activated charcoal, and a number of preconcentration procedures are based on recovery of sorbed material from this solid. Greater selectivity may be achieved through sorption on suspensions of organic reagents. Other published procedures support specific reagents on solid matrices such as cellulose or polyurethane foam. In many respects these loaded materials resemble chelating ion exchangers.

The amount of material recovered from a solid collector is not always consistent and this detracts from its popularity as a preconcentration method. In fact, in trace analysis, concern with this phenomena is usually restricted to estimations of the amount of loss which might occur through sorption of solutes of interest onto the walls of containing vessels. Where suspended solids are present (e.g., in natural waters), the amount of labile solute present also becomes a function of the amount and type of suspended matter.

II. MASKING

Masking is a process for eliminating interference effects based on transforming the chemical nature of substances so that certain of their usual reactions are prevented. The aim is to avoid the need for actual physical separation of the interferants or their reaction products from the assay solution.

An example of masking is the use of a hydroxy carboxylic acid (e.g., citric or tartaric) to prevent the precipitation of iron(III) (as the hydrous oxide) on the addition of ammonia. Similarly, by adding cyanide ions to a solution containing copper, zinc, cadmium, magnesium, and calcium salts, titration with EDTA solution results in reaction with the calcium and magnesium ions only; the heavy metals are masked in the form of cyano complexes.

Masking involves the selection of conditions which will ensure that one chemical reaction (e.g., complex formation) predominates over a competing reaction (e.g., precipitation). Prediction of the optimum conditions requires consideration of all the chemical equilibrium situations which may develop in the assay solution.

Consider the example of the masking of iron(III). The principal reaction (to be prevented) is precipitation: $Fe^{3+} + 3OH^- \rightleftarrows Fe(OH)_3$; and from the solubility product principal it can be proposed that a precipitate will be formed only if $[Fe^{3+}][OH^-]^3 > K_{So}$, i.e., $>10^{-38}$.

In the presence of ammonia and ammonium salts the pH of the solution (i.e., $-\log [H^+]$) is controlled by the relationship $H^+ + NH_3 \rightleftarrows NH_4^+$; for which $K_a = [H^+] [NH_3]/[NH_4^+] \simeq 10^{-9}$ or $[H^+] = 10^{-9} [NH_4^+]/[NH_3]$. In the absence of a large excess of ammonium salts, the $[H^+]$ is thus generally $< 10^{-7}$ (i.e., $[OH^-] > 10^{-7}$) and hydrous iron oxide can be expected to precipitate unless $[Fe^{3+}]$ is reduced to values of $< 10^{-17}$ M (pH 7), or $< 10^{-28}$ (pH 8), etc.

To achieve values of $[Fe^{3+}]$ which are $< 10^{-17}$, the iron(III) has to be converted into a stable complex ion.

If the reagent selected was ethylenediaminetetraacetic acid (abbreviated to H_4Y) the reactions of interest would be

$$Fe^{3+} + Y^{4-} \rightleftarrows FeY^- ; \quad K_c \simeq 10^{-25}$$

$$Y^{4-} + xH^+ \rightarrow HY^{3-} + H_2Y^{2-} + H_3Y^- + H_4Y$$

By appropriate manipulation of the equilibrium equations appropriate to the stepwise addition of protons to Y^{4-}, one can derive the relationship $[Y]_T = \alpha[Y^{4-}]$, where $[Y]_T$ is total excess ligand (in all forms), and

$$\alpha = 1 + \frac{[H^+]}{K_4} + \frac{[H^+]^2}{K_3K_4} + \frac{[H^+]^3}{K_2K_3K_4} + \frac{[H^+]^4}{K_1K_2K_3K_4}$$

Since K_1, K_2, K_3, and K_4 are known, one can evaluate α as a function of $[H^+]$.

Substitution of $[Y]_T/\alpha$ for $[Y^4]$ in the metal complex stability equation yields the relationship

$$\frac{[FeY^-]}{[Fe^{3+}][Y]_T} = \frac{K_c}{\alpha} = K_{eff}$$

To achieve values of $[Fe^{3+}]$ which are $\ll 10^{-17}$ [i.e., where $[FeY^-] \simeq$ total iron(III) content] one obviously must minimize $[H^+]$ and maximize $[Y]_T$.

Increasing the pH increases the fraction of total EDTA present as Y^{4-}, but at the same time favors precipitation. For example, it can be shown that with a total EDTA concentration of 0.1 M, the values of $[Fe^{3+}]$ at pH values of 8, 9, and 10 are respectively approximately 10^{-23}, 10^{-24}, and 10^{-25}. To form a precipitate at these pH values, $[Fe^{3+}]$ should be $> 10^{-20}$, 10^{-23}, or 10^{-26} M respectively. These results indicate that some precipitation would occur in solutions of pH > 10; i.e., masking fails at pH > 10.

Simple calculations such as these indicate the feasibility of using certain reactions for masking purposes. More reliable predictions are possible if corrections are made for the influence of ionic strength on activity coefficients and for the effect of temperature on the magnitude of equilibrium constants. In the absence of this information, the effectiveness of masking has to be determined by experiment.

The factors to be considered in a masking reaction include definition of the nature of the principal reaction (to be prevented); the number of masking reagents required to react with all interfering species; the

amount of principal reagent and masking agent to be used; the pH of
the solution; the rate of the various reactions; the effect of changes
of oxidation state; the nature of solutes and ionic strength in the solvent medium; and the effect of temperature.

While the number of reactions which could be adopted for masking
is theoretically quite large, satisfactory results have been achieved
using a comparatively small list.   Some typical reagents are listed
in Table 8.1.   While each reagent is shown to react with an appreciable
number of elements, it does not mean that all of the elements are satisfactorily masked by this reagent in a series of different determinations. It may also be noted that in order to mask various combinations
of elements, several masking agents may have to be added to the base
solution.

For many purposes masking is a very convenient method of over-
coming interference effects (e.g., in colorimetry, solvent extraction,
titration with complexing agents such as EDTA, etc.). However, where
the number of elements to be masked is large, or where the calculated
$[M^{n+}]$ values for the principal and masking reaction are similar in
magnitude, it is often advisable to adopt alternative methods of separation or analytical determination.

III. SELECTION OF METHODS

Analytical procedures have been developed which utilize almost all the
known chemical and physical properties of atomic and molecular species.
The various techniques differ in terms of selectivity, sensitivity,
simplicity, equipment requirements, or fields of application.

The preceding chapters contain brief summaries of the principles
of a selection of the more widely applied techniques. A comprehensive
treatment of the range available would need to include nearly four times
as many topics.

By adjustment of conditions, most techniques can be used for a wide
range of analyses, but in order to broaden the range of applications one
usually has to increase the complexity of preliminary steps or the time
required, and accept a decrease in overall accuracy. For each technique there is normally a restricted area in which the technique is
preferable to all others.

The number of options available to a chemist when faced with the
problem of selecting an appropriate method for a new situation can be
bewildering. Selection of the best method can require the knowledge,
skill, and wisdom which comes only with years of experience.

Table 8.1

Common Masking Agents

Masking agent	Element masked
Citrate or tartrate	Al, Ba, Be, Bi, Ca, Cd, Ce, Co, Cr, Cu, Fe, Hg, Mn, Mo, Ni, Pb, Sb, Sn, Sr, Th, U, Zn.
EDTA[a]	Al, Ba, Bi, Ca, Cd, Ce, Co, Cr, Cu, Fe, Hg, Mg, Mn, Mo, Ni, Pb, Sb, Sr, Th, V, Zn.
F^-	Al, B, Be, Ca, Ce, Cr, Fe, Th, Ti, U, V, W, Zn, Ge, Mg, Mn, Mo, Nb, Ni, Sn.
CN^-	Ag, Au, Cd, Co, Cu, Fe, Hg, Mn, Ni, V, Zn.
Oxalate	Al, Fe, Ge, Mg, Mn, Mo, Nb, Sn, U.
CNS^-	Ag, Cd, Co, Cu, Fe, Mo, Ni, W, Zn.
BAL[b]	Al, As, Bi, Cd, Co, Cu, Fe, Hg, Mn, Pb, Sb, Sn, Zn.
NH_3	Ag, Co, Cu, Fe, Ni, Zn.
H_2O_2	Co, Mo, Nb, Ti, U, V, W.

[a] EDTA = ethylenediaminetetraacetic acid.

[b] BAL = 2,3-dimercapto-1-propanol.

In this short introduction, the aim is solely to indicate the logical processes or steps involved.

The prime prerequisite is definition of the analytical problem. The checklist of factors to be considered includes items such as the following

1. Nature of the species to be determined, e.g., elemental, molecular, or functional group.

2. Concentration range to be studied, e.g., μg/liter, μg/ml, mg/ml, etc.

3. Accuracy desired, time available for each analysis.

4. Frequency of sample delivery, volume of sample flow.

5. Restrictions due to physical or chemical properties of the samples, e.g., corrosive gases, radioactive solids.

6. Potential interference effects attributable to the matrix.

7. Analytical facilities available, i.e., quality and quantity of specialized equipment on hand.

8. Environmental limitations, e.g., are analyses to be performed in an industrial plant, out in the field, or in a modern laboratory ?

9. Quality of trained personnel available.

10. Economic justification, i.e., does the value of the decision arising from the analysis justify the cost per determination?

Once the analytic problem has been defined, the selection of provisional methods can begin. This process is greatly facilitated by judicious use of the tabulations and compilations available in the literature. For example, in the Handbook of Analytical Chemistry the material is organized in terms of specific techniques and in terms of analysis of different materials. The Handbook provides summaries of alternative procedures and through references provides a guide to original literature.

If methods of high reliability and accuracy are desired, one is well advised to consult the publications of national organizations such as the American Society for Testing Materials, the British Standards Association, etc. For routine determinations, monographs on individual topics such as "The Analysis of Iron and Steel" or "The Quantitative Analysis of Drugs" provide a range of options.

Other sources of potential methods include multivolume reference works, "Advances in...." series, review articles, analytical abstracts, and analytical journals.

Whatever the sources used, a list of procedures which may be suitable for a given project should be compiled. The next step is evaluation of the merits of the alternative procedures.

The alternatives should be considered in terms of questions such as:

1. Does the procedure meet all, most, or just some of the conditions outlined in the problem definition?

2. What is the maximum accuracy which might be achieved? What special equipment is required? How skilled an analyst is required? How long should the procedure take?

3. What species interfere in this particular method? Are they likely to be in the sample matrix? What steps are recommended to minimize the effect of interferents?

4. Are all of the alternative procedures subject to the same degree of interference? Which is the simpler procedure?

From the possibilities, one technique or procedure may appear preferable to all others, and the final step is testing of the proposed procedure. This testing is performed by using the procedure to analyze standard samples (of composition similar to the future unknowns) and subsequently using the data to evaluate accuracy and precision limits. Only after the method has been proved reliable should it be applied in routine studies.

Any person engaged in quantitative analysis must be continually critical of proposed procedures, conscientious in testing and evaluating, and ever ready to modify steps to allow for new interference patterns.

The evaluation of pollution poses some entirely new challenges, in seeking answers to these challenges the knowledge and resources of modern analytical chemistry can be an invaluable aid.

In conclusion, one may remark that there are many problems and many steps in the M.I.L.E. which lies between the starting point (S) in a chemical analysis and the final data and subsequent decisions (D) derived from the experimental study.

Selection of sampling procedure
S   Storage problems
Sample preparation difficulties.

Method selection: definition of problem, critical evaluation
M   of alternatives, testing with standards;
Modification due to matrix effects.

I   Interference effects: identification of species or phenomena
which can lead to spurious results.

Link reactions required to achieve effective separation of
L   components, or elimination of interference effects.

Evaluation of the amount of element or molecular species of
E   interest using a technique appropriate to the concentration
range being studied.

Decisions based on chemical data. The most logical decision
D   is limited in quality by the validity of the results obtained in
the experimental studies.

FURTHER READING ON ANALYTICAL CHEMISTRY

It is hoped that this introduction to pollution evaluation has raised
the enthusiasm of the reader to a level where he or she wishes to know
more about the techniques used to obtain accurate and valid quantitative
data.

If the interest is primarily in a particular technique, one may proceed directly to appropriate monographs, most top-class libraries carry
a broad selection of such books.

On the other hand, it is recommended that a sound broad background
be first developed through reading one or more of the larger general
texts prepared for use in tertiary institutions. A list of some of the
books released in the last decade is given below.

After laying suitable foundations, one may expand in depth in any
desired direction, through perusal of specialist books, comprehensive
treatises, selective reviews, and journal publications.

GENERAL TEXTS ON QUANTITATIVE CHEMICAL ANALYSIS

Ayres, G. H.: "Quantitative Chemical Analysis," 2nd ed., Harper,
N.Y., 1968.

Day, R. A., and Underwood, A. L.: "Quantitative Analysis," 2nd ed.,
Prentice-Hall, Englewood Cliffs, N.Y., 1967.

Dick, J. G.: "Analytical Chemistry," McGraw Hill, N.Y., 1973.

Ewing, G. W.: "Instrumental Methods of Chemical Analysis," 4th ed., McGraw-Hill, N.Y., 1975.

Fischer, R. B., and Peters, D. G.: "Quantitative Chemical Analysis," 3rd ed., Saunders, Philadelphia, 1968.

Fritz, J. S., and Schenk, G. H.: "Quantitative Analytical Chemistry," Allyn and Bacon, Boston, 1966.

Flaschka, H. A., Barnard, A. J., and Sturrock, P. E.: "Quantitative Analytical Chemistry," Barnes and Noble, N.Y., 1968.

Grunwald, E., and Kirschenbaum, L. J.: "Introduction to Quantitative Chemical Analysis," Prentice-Hall, Englewood Cliffs, N.J., 1972.

Kolthoff, I. M., Sandell, E. B., and Meehan, F. J.: "Quantitative Chemical Analysis," 4th ed., Macmillan, N.Y., 1969.

Pecsok, R. L., and Shields, L. D.: "Modern Methods of Chemical Analysis," Wiley, N.Y., 1968.

Pickering, W. F.: "Fundamental Principles of Chemical Analysis," Elsevier, Amsterdam, 1966.

Pickering, W. F.: "Modern Analytical Chemistry," Dekker, N.Y., 1971.

Pietrzyk, D. J., and Frank, C. W.: "Analytical Chemistry, An Introduction," Academic Press, N.Y., 1974.

Siggia, S.: "Survey of Analytical Chemistry," McGraw-Hill, N.Y., 1968.

Skoog, D. A., and West, D. M.: "Fundamentals of Analytical Chemistry," 2nd ed., Holt, N.Y., 1969.

INDEX

A

Absolute error, 15
Absorbance, 54, 90, 94, 116
Absorptimetry, 48, 50, 90, 116-120, 165-168, 172, 183 (see also Colorimetric methods)
Absorptiometer unit, 50, 58
Absorption cell, 50
Absorption column (counter current), 48, 50
Absorption of radiation:
  infrared, 43, 54, 91-95, 152, 169, 170
  uv-visible:
    by atoms, 83-85, 120, 138
    by molecules, 84, 88-95, 109, 116
Absorption spectra:
  infrared, 53, 54, 92, 94, 116
  uv-visible, 89, 90, 116
Absorption spectrophotometry, 116-120 (see also Absorptimetry)
Accuracy, 1, 14-22, 24, 59
Acid-base indicators, 70, 71
Acid-base titrations, 66-73, 80, 178
Acid dissociation constant, 60, 67-70, 164, 182
Acid dissolution (soils, plants), 162

AC Polarography, 126
Activated charcoal, 109, 165, 180
Activity coefficient, 59, 130, 145, 182
Activity (radioactive), 155, 157
Activity of species, 59, 69, 74, 130, 132, 145
Additivity of absorbance, 118
Adsorption:
  coefficients, 9, 10
  of copper species, 9
  of gases, 32, 144, 148, 179
  isotherms, 9, 62, 179, 180
  losses, 28, 103, 105
  on precipitates, 59, 61, 62
  sites, 9, 180
  tubes (vapor tests), 32
Adsorptive capacity, 7
Aerosols, 83, 139, 141
Aerotoxicants, 51, 53
Affinity (for adsorbents) 9, 180
Albuminoid ammonia, 114
Alcohol vapor (in breath), 32, 56
Aldehydes (in air), 46, 47, 51
Ammonia (in water) 113, 114
Analytical challenges, 4, 185
Analytical texts, 187, 188
Anion exchangers, 176-179
Anodic stripping voltammetry, 99, 127, 128
Arcs, 141, 155
Aromatic compounds (in air), 46, 51